Praise for *Growing Papaya Trees*

"In *Growing Papaya Trees*, Jessica Hernandez makes the case that Indigenous science is the best way to challenge the current climate crisis. Through beautiful prose and storytelling based on her own Maya Ch'orti' and Zapotec people's experiences with settler colonialism, she argues that we need collaboration and solidarity from all Indigenous communities, whether on their homelands or displaced into diaspora, and it is through our collective strength that we can meet the challenges of our times. If you want to learn about the importance of Indigenous science and papaya trees to address the climate crisis produced by colonialism, this is a must-read."

—KYLE T. MAYS (BLACK AND SAGINAW CHIPPEWA), author of
An Afro-Indigenous History of the United States

"A critical and timely contribution to the discourse on how Indigenous land-based knowledge, relationships, and stewardship can address climate displacement. Hernandez weaves together realities of Indigenous displacement, marginalization, and ongoing impacts of colonialism with stories and teachings rooted in Indigenous resilience and solidarity. Her deep connection to and love for ancestral homelands infuses the book with an intimacy that is rare and powerful. The teachings embedded in the book are not only important for other Indigenous people and scholars, but are also critical for everyone who cares about our shared histories and futures."

—STYAWAT / DR. LEIGH JOSEPH, author of *Held by the Land*

"With *Growing Papaya Trees,* Dr. Hernandez breaks open discussions of diasporic Indigenous identity, shared histories, and climate displacement. Drawing from personal narratives and compelling research, Hernandez has provided an urgent second book that calls for both a keen eye on the past and solidarity in the present."

—KINSALE DRAKE, author of *The Sky Was Once a Dark Blanket*

"Transported by stories that remind us that the Land remembers, Dr. Hernandez weaves tales that carry ancestral wisdom—guiding us toward collective survival, healing, and belonging. This book is a prayer, a balm, and a call to return to each other and to the earth. At a time when our communities are rising against violence and erasure, these stories of migration, sacredness, and resistance are more vital than ever."

—CÉLINE SEMAAN, author of *A Woman Is a School*

Praise for *Fresh Banana Leaves*

"Westerners, [Dr. Hernandez] writes, fall short on including Indigenous people in environmental dialogues and deny them the social and economic resources necessary to recover from 'land theft, cultural loss, and genocide' and to prepare for the future effects of climate change."

—*PUBLISHERS WEEKLY*

"In *Fresh Banana Leaves*, Jessica Hernandez weaves personal, historical, and environmental narratives to offer us a passionate and powerful call to increase our awareness and to take responsibility for caring for Mother Earth."

—EMIL' KEME (K'ICHE' MAYA NATION), member of the Ixbalamke Junajpu Winaq' Collective

"A groundbreaking book that busts existing frameworks about how we think about Indigeneity, science, and environmental policy. A must-read for practitioners and theorists alike."

—SANDY GRANDE, professor of political science and Native American and Indigenous studies, University of Connecticut

"Inspiring and sobering, philosophically powerful and practically grounded, this book weaves together storytelling, razor-sharp critiques of oppression, and liberatory pathways for how we can achieve transformation in solidarity. Dr. Hernandez offers the instructions so many environmental protectors and conservationists need to know."

—KYLE WHYTE, George Willis Pack Professor, School for Environment and Sustainability, University of Michigan

Growing Papaya Trees

ALSO BY JESSICA HERNANDEZ

Fresh Banana Leaves

GROWING PAPAYA TREES

NURTURING INDIGENOUS ROOTS DURING CLIMATE DISPLACEMENT

JESSICA HERNANDEZ, PhD

North Atlantic Books
Huichin, unceded Ohlone land
Berkeley, California

North Atlantic Books
Huichin, unceded Ohlone land
2526 Martin Luther King Jr Way
Berkeley, CA 94704 USA
www.northatlanticbooks.com

Cover art © The Naturalist via Getty Images
Cover design by Jasmine Hromjak
Book design by Happenstance Type-O-Rama

Printed in Canada

Growing Papaya Trees: Nurturing Indigenous Roots During Climate Displacement is sponsored and published by North Atlantic Books, an educational nonprofit that collaborates with partners to develop cross-cultural perspectives; nurture holistic views of art, science, the humanities, and healing; and seed personal and global transformation by publishing work on the relationship of body, spirit, and nature.

North Atlantic Books's publications are distributed to the US trade and internationally by Penguin Random House Publisher Services. For further information, visit our website at www.northatlanticbooks.com.

The authorized representative in the EU for product safety and compliance is Eucomply OÜ, Pärnu mnt 139b-14, 11317 Tallinn, Estonia, hello@eucompliancepartner.com, +33757690241.

Library of Congress Cataloging-in-Publication data is available from the publisher upon request.
ISBN: 979-8-88984-097-8 (paperback)
ISBN: 979-8-88984-098-5 (epub)

The interior of this book is printed on 100 percent recycled paper, and the cover is printed on material from well-managed forests.

1 2 3 4 5 6 7 8 9 FRIESENS 30 29 28 27 26 25

This book is dedicated to my ancestors, whose stories went untold and whose stories were denied the right to be told.

CONTENTS

A Love Letter to Our Ancestral Lands

Dear Ancestral Lands,

    We yearn to return and reclaim our ancestral homelands, but until then, this letter serves as a testament to our unwavering connection to our Lands. Our connection to our Lands is a bond that cannot be adequately expressed in words. It is a bond forged through generations, through stories passed down from our ancestors, and through countless experiences shared with our Lands. It is a bond that transcends time and space, linking us to the Lands in ways that feel both familiar and foreign.

    Our Lands will never forget us. Our Lands will always honor our cries, laughter, and presence. Through our Lands, our stories will be passed on for generations to come. Our Lands will remember our love and respect, which will be our legacy, as we do not truly own our Lands. We cannot take our Lands with us into the afterlife. As Indigenous Peoples, we are a part of our Lands, and our Lands are a part of us. Our Lands will remember our connection and unity. Our relationship with our Lands is sacred and will live on in our hearts and memories. Our Lands will always hold a special place in our souls. Even though we may have to leave our Lands due to forced displacement, our Lands never leave us behind. Our Lands are our homes and our forever caregivers.

    Our roots are nourished and maintained by our Lands, even when our roots extend to unfamiliar places we call home away from home. We eventually return to our roots, which are nourished and maintained by our Lands. This is the cycle of life for us—we come from our Lands and return to our Lands. As we pass on, our Lands care for us, they provide new life and nourishment. We may die alone, but we do not rest alone. With our ancestors' spirits, stories, and footsteps, our Lands accompany us on our way to eternal peace. Our Lands are more than physical objects or property. Our Lands are our mother and spirit, and we cannot own our mother or sell spirits.

We are often silenced because our ways of living and being are in opposition to settler governments. We are our Lands, and thus, even when someone else claims to own our Lands, the claimer cannot own us. Our Lands are who we are, who we become, and what we hold, as Indigenous Peoples, dear in our hearts. Our Lands are our oldest cultural heirlooms, which we will protect for generations to come. Although we may physically leave our Lands, our Lands will never leave us. This love between our Lands and us is ingrained in our collective memory as Indigenous Peoples. We must recognize that our Lands are the foundation of our identity and an integral part of our culture and traditions.

As Indigenous Peoples, we have always understood the importance of environmental stewardship. We have lived in harmony with our Lands for centuries, respecting our Lands' rhythms and cycles. We understand the interconnectedness of all living beings. We must honor our Lands, we must protect our Lands, and we must fight for our Lands.

Precious Lands, we look forward to the day when we can again walk on your sacred grounds and feel your healing embrace.

Sincerely,
Displaced Indigenous Peoples

PROLOGUE

The men who walked in darkness, fleeing nightmares that disallowed them from dreaming in their homelands. The women who braided their hair to avoid being caught by the winds whispering their futures. The children whose cries were silenced by the storms. These are the people who are forced to uproot themselves in search of Lands free from chaos, free from natural disasters, and free from persecution for being Indigenous. They are looking for Lands replenished with opportunities that will allow them to survive and support their loved ones back home. Yet, little do they know, such Lands do not exist as the entire world is rampant with colonialism and systems that oppress Indigenous Peoples. However, for many Indigenous Peoples from the Global South, this is the only opportunity or option they have. They are displaced from their ancestral Lands, ripped from their identities, and hidden in plain sight. This is because their existence, voices, and stories are too complex to encapsulate within the boxes of racial categories created by a foreign country. They are the guardians of their stories, the cultural bearers of their traditions, and the language teachers of their mother tongues. Still, too often, they are forced to assimilate and hide in disguises to avoid being detained by immigration services. If caught or found, they would be forced back to the Lands they fled from. They disguise themselves not out of shame but for survival. They have no choice. They must do whatever it takes to find a country that offers them the right to survive and thrive, even though such a country is impossible to find as it does not exist. In every country they seek refuge in, they will face discrimination and hostility for being outsiders. Fear of deportation and the struggle to integrate into the country they desperately sought will add to their challenges. Despite their hopes for a better life, they encounter barriers that make it difficult to truly belong anywhere.

When we read about immigration, we only find stories of economic prosperity and opportunities. We rarely hear about immigrants' struggles with language barriers, cultural shock, and discrimination. We don't hear about immigrants who are exploited and taken advantage of or lose their lives as they migrate. We never see the tears that streamed down their cheeks as they kissed their families goodbye and received their blessings. A misleading picture of immigrant experience is painted by the media when stories are made up to suit a political agenda. Their focus is on what is convenient at that time to portray a certain picture while neglecting the complexities and hardships immigrants face. This selective storytelling can shape public perception and policy, ignoring the need for support and reform. The narrative reinforces the idea that immigrants are criminals and have broken the law, overshadowing the systemic barriers they face. By focusing only on legalities, the media does not address the inequalities and struggles of those who do not fit the idealized vision of the American Dream. The distorted portrayal leads to a misunderstanding of the true migrant experience and the urgent need for comprehensive reform as a result.

Their criminalization hides their families who understood the dangers they were about to face when they made the decision to leave their homes. Unfortunately, there are a fortunate few, since many people die on such journeys, leaving behind spirits that cannot find eternal rest. Sadly, these spirits cannot transcend to meet their ancestors after a traumatic end leaves them caught between the spirit and living worlds. They are trapped wishing they had fulfilled their dream of finding such Lands that would allow them to live happily ever after.

Little is shared about the dire circumstances and desperation that drove them to make such a perilous journey. As a result, the human aspect of their plight remains largely invisible, overshadowed by the sensational tales of their restless spirits. This is why goodbyes are difficult to endure. All of us fear never seeing each other again. Sometimes, death can become us. Throughout our journey of displacement, Indigenous Peoples face hazardous conditions such as dangerous terrain, extreme weather, and treacherous water crossings.[1] Additionally, we may fall victim to human trafficking, violence from smugglers, and exploitation by those who prey on our vulnerability. Even if we survive these perils, we often experience severe health risks due to a lack of food, water, and medical care. Despite the risks we take to find opportunities, we as Indigenous Peoples are considered criminals, as if this "crime" was easy to commit. Our People never want to leave our homelands, but instead of hearing of the longing many displaced Indigenous Peoples have toward returning to their Lands, we often hear xenophobic and anti-immigrant narratives. They say, "Indigenous Peoples must return to their Lands"; however, our Lands are no longer livable. The waters are polluted, the forests have been clear-cut, and the soil is exhausted from decades of exploitation. Climate change has brought extreme weather, making it impossible to sustain traditional ways of living. In such conditions, our communities have no choice but to seek refuge elsewhere.

During these journeys, we don't know who to trust. In many cases, the people who smuggle Indigenous Peoples are also the people who sell them out. They are known as coyotes. The label *coyote* is often used to describe people who smuggle migrants into the country. They are the ones who promise to guide people across the border; however, coyotes exploit their desperation by charging them exorbitant fees and even abandoning them in dangerous conditions. In Indigenous stories, coyotes are sometimes represented as tricksters. Coyotes, the animals, are known for their cunning, resourcefulness, and adaptability, which make them perfect sly figures. They often outsmart their prey and can navigate harsh environments easily. In Indigenous folk tales, these traits allow coyotes to outwit and take advantage of others through cunning. Similarly, coyotes are untrustworthy because migrants do not realize they are their prey. Even when migrants are aware, they may feel they have no other option. The impact of coyotes on migrant communities is profound and often devastating.[2] They exploit the vulnerabilities of those seeking a better life, leading to situations of financial ruin, physical danger, and emotional trauma. This exploitation erodes trust within the communities, as individuals fear that those meant to help them might instead lead them into harm's way. The emotional toll on migrants as they navigate a constant state of fear and uncertainty is immense. The betrayal by those who promised to guide them safely can lead to deep psychological scars and a pervasive sense of mistrust. This emotional burden compounds the already challenging journey, making the pursuit of a better life fraught with despair and anxiety.

Many of us depart from our Lands holding hope in our hearts while feeling this is our only option. We leave our families knowing far too well that many of us will never see them again. We leave our homes knowing we will never return. This is the story of many displaced Indigenous Peoples. Along with mistrust and fear, the toll of such displacement leaves individuals with a deep sense of loss and longing, an enduring ache that can never truly be healed.

Our parents and grandparents never wanted us to leave our Lands. Our parents and grandparents never wanted us to leave our families. Still, they understand staying means economic hardships, lack of educational opportunities, limited access to healthcare, and few prospects for a stable career. They know we have no choice, so, with broken hearts, we say our goodbyes. As difficult as it is to leave, these sacrifices are made in hopes of a brighter and more secure future. Climate change leads to unpredictable weather patterns, dwindling natural resources, and increasing natural disasters contributing to the struggle of living off our Lands.[3] These environmental changes exacerbate these challenges forcing Indigenous Peoples out.

Indigenous Peoples endure a complex reality. A reality that is too often romanticized for the media and internet clicks but never truly acknowledged. Beneath the fantasy are the hidden stories of rape, violence, and death that many Indigenous Peoples experience or witness. However, you will never see their faces, read their stories, or learn from their experiences. They are made invisible by the American Dream that either romanticizes or criminalizes migration. Migration is ugly when it involves people, but the migration of butterflies or other animal species is defined as a natural wonder. This juxtaposition highlights the selective empathy toward certain types of movement while ignoring the harsh realities faced by human migrants. It is a stark reminder that our perceptions are shaped by narratives that often dehumanize those most in need of compassion.

Human migration is criminalized through restrictive policies, detentions, and deportations that strip individuals of their dignity and rights.[4] Migrants are often portrayed as threats or burdens rather than as human beings seeking safety from their current, harsh conditions. This selective empathy underscores the need for a profound shift in how society views and treats those who cross man-made borders. In ancestral times, borders as we know them did not exist, and movement was a natural part of life. People roamed freely, guided by the seasons, the availability of resources, and the need for survival. This historical context emphasizes that the modern concept of borders is a social construct that serves to divide and dehumanize Indigenous Peoples of this continent. Our parents, now in their old age, recall the beautiful childhood memories they once shared with their loved ones and their Lands. Their childhood stories are filled with their heritage and the natural environment that defined their way of life. Those recollections shaped their identities and served as reminders of the community and culture that thrived before displacement. The American Dream is nothing but a myth for many of us; we know this narrative was not created for displaced Indigenous Peoples.

If they do seek the American Dream, Indigenous Peoples face significant obstacles such as financial instability, lack of education access, and systemic inequalities. Behind the so-called American Dream lies the reality that if someone, no matter their nationality, can embody whiteness or proximity to it within the settler boundaries, they can benefit from such a narrative. Like people in other marginalized groups, many displaced Indigenous Peoples from the Global South do not benefit from whiteness. The narrative of the American Dream, therefore, remains elusive for those who do not fit the mold of the dominant cultural and racial paradigm. Indigenous Peoples do not fit this mold, and thus, many of us suffer unfair labor practices, especially in agriculture. These practices include exploitative wages, unsafe working conditions, and a lack of job security. Despite our dedication to work, prosperity and upward mobility remain out of reach for most of our people. Many farmworkers are displaced Indigenous Peoples who face horrendous working conditions that often go unaddressed. They endure long hours in extreme weather, can take only minimal breaks, and face exposure to harmful chemicals without adequate protection. This exploitation highlights the harsh realities those who labor to sustain the nation's food supply suffer. Back home, we know the job conditions that await in the United States. Despite this knowledge, many of us still choose to come to the Land we call El Norte (the north), because we can no longer live off the Lands of our lineages. Unfortunately, the reality we face often falls short of the dreams we carry with us. The harsh reality is that many find themselves trapped in a cycle of exploitation and marginalization.

El Norte, the first words many of us remember learning as displaced Indigenous Peoples from south of the border. They are the words that start every chapter of our lives after moving away from our Lands, away from our grandparents, and away from our communities' apex. They are the words we fear but must accept. Sadly, they are the only words our relatives can whisper in hopeless times. Time and time again, we cross borders with just memories no immigration agent can take from us. Our journeys to El Norte are not fairytales; they do not end with "happily ever after." These paths are found in every household of displaced Indigenous families, even if they would not dare to share them publicly. Testimonios about

El Norte are not told to others but rather kept as family secrets shared only in the darkness of night. No matter how hard we try to silence these stories, the spirits refuse to stay hidden. Despite the risk of persecution, the energies are always unleashed. This is a reminder that no matter how hard we try to repress our stories and experiences, they cannot be contained. Our histories and experiences will always be a part of us.

Our stories help us preserve our cultural identity. This is significant to us as displaced Indigenous Peoples because they serve as a link to our ancestors and traditions. These understandings ground us in a sense of belonging and purpose. They allow us to pass down values, wisdom, and history to future generations, ensuring that our heritage is not lost in the face of adversity.

By embracing and sharing our stories we promote diversity and understanding. We must find ways to share our truth with the world, a world that often homogenizes and marginalizes. Our stories remind us of our shared humanity and the power in collective action. This is why Indigenous solidarity is critical, even when we come from different settler states. By uniting and supporting one another, we can amplify our voices and advocate more effectively for our rights. We, as displaced Indigenous Peoples, are not here to steal anyone's Lands; rather, we seek recognition and respect for our own. We aim to coexist harmoniously while preserving our cultural heritage and ancestral connections through respect and empowerment. We advocate for justice and protection of our rights, not at the expense of others but through mutual understanding and collaboration.

I, as an Indigenous woman from the Global South, am not here to compete with anyone else. Instead, I aspire to contribute my unique perspective and experiences to enrich our collective knowledge. Our shared goal is to uplift and support each other in our efforts to preserve and protect our cultural heritage. One of the reasons why I left full-time academia was the enigma of feeling as though there was not enough room for more than a few Indigenous scholars and professors. Academia is a settler colonial project that often pits marginalized voices against each other rather than fostering genuine collaboration and solidarity. This environment can be detrimental to our collective efforts and organizations. Throughout my academic journey, I met amazing Indigenous scholars and professors who fostered a sense of community and belonging instead of competition. These individuals have been instrumental in shaping my perspective and encouraging me to pursue my passion for cultural preservation. Their unwavering support and mentorship have shown me that solidarity is possible, even within challenging environments. I share these experiences because, as displaced Indigenous Peoples, it is the xenophobic ideologies that paint us as adversaries rather than collaborators. The rhetoric that "Immigrants are here to steal our jobs" is harmful and divisive, and it ignores the rich contributions that displaced communities bring to society. By challenging these narratives and fostering unity, we can work toward a more inclusive and respectful world for all.

Just as papaya trees thrive and survive in harsh conditions, we are here to thrive and survive as well. Through storytelling, we share experiences and lessons that strengthen our community and help us navigate challenges. By listening to each other's stories, we gain insight and inspiration to persevere. Together, we can cultivate resilience and support one another in overcoming the obstacles we encounter.

Papaya trees hold profound significance in many Indigenous cultures, symbolizing resilience, sustenance, and connection to the Land. These trees provide essential nourishment but also represent deep-rooted traditions and wisdom passed down through generations. Papaya trees aid in community resilience by providing a reliable source of food and nutrition, which is crucial during times of scarcity. Their ability to thrive in diverse climates makes them a sustainable crop, ensuring food security for Indigenous communities. The act of cultivating these trees fosters a sense of unity and collective effort.

By nurturing papaya trees, Indigenous communities honor their heritage and strengthen their bond with nature, even in the face of displacement and environmental challenges. We do not believe in individuality or competition with one another. Papaya trees grow rapidly and bear fruit within months of planting. Their ability to adapt and thrive in a variety of climates and soil conditions symbolizes resilience, perseverance, and strength. Indigenous Peoples are like papaya trees. We are the trees that the displaced roots grew, and we are here to advocate for our experiences, existences, and resistances.

Sharing these narratives helps preserve our cultural identity, honors the enduring spirit and resourcefulness of displaced Indigenous Peoples, and empowers the next generation to continue the legacy of resilience. Through our words, the youth can feel a sense of belonging despite the never-ending longing to return to our ancestral Lands.

It is rare to see stories told by displaced Indigenous Peoples from the Global South, so it is up to those of us who can share our histories to do so. We must tell our stories to ensure that our traditions and wisdom are not lost but instead thrive despite the challenges we face. This is how we nourish our roots and heal our displacement wounds. *Growing Papaya Trees: Nurturing Indigenous Roots During Climate Displacement* is an homage to our ancestors who were denied the right to tell their own stories. Their truths will not remain untold.

INTRODUCTION

Immigrant is a word we have become too familiar with, especially once we have been displaced in Lands not our own, Lands that have now become settler territories for all Global North nations. This label often comes with illegal or undocumented status. This adds another layer of unwelcomeness and inhospitality we are greeted with for simply trying to survive. This label can evoke feelings of isolation, fear, and rejection, making it difficult to find a sense of belonging. It often leads to discrimination and prejudice, further exacerbating the emotional toll on individuals striving to build a new life. The constant struggle to prove one's worth and humanity becomes an exhausting daily battle. Especially when our ancestral collective memories point to a reality that we too often fail to acknowledge we once engaged in cultural exchanges before these manmade borders existed. The xenophobia found, even within Indigenous communities, contrasts sharply with our shared history of mutual respect and cooperation. Recognizing this can help us reframe our current interactions and work toward more inclusive and understanding transnational Indigenous solidarity.

Within this discourse, we as displaced Indigenous Peoples must respectfully recognize that Indigenous sovereignty and presence precolonization existed and continue throughout history. This affirmation reminds us of Indigenous communities' resilience and enduring strength. By honoring this sovereignty, we can forge solidarity and mutual support and foster a future where Indigenous voices are heard and appreciated across all borders. Indigenous histories are often erased by the narratives created by colonizers to obscure the atrocities they committed against us and our ancestors. It is unfortunately common for Western colonial projects to erase Indigenous histories as a way to conceal the wrongs inflicted upon us and our ancestors. This erasure is so thorough that Indigenous perspectives are often absent from the history books used in our early education. As a result, many people grow up with a limited understanding of the true impact of colonization. This lack of representation hinders efforts to acknowledge and address historical injustices. Indigenous communities across the world face similar challenges to cultural erasure and misrepresentation. This shared experience unites Indigenous Peoples, reinforcing solidarity and collective action. We must do this by respecting one another's sovereignty and autonomy.

As Indigenous Peoples from the Global South, we can only respect our northern Indigenous relatives and stand in solidarity with them. We should recognize their sovereignty and right to self-determination and autonomy. We must also recognize our responsibility to help repair the wrongs committed against all Indigenous Peoples who have been displaced from their Lands. It is the way of life our elders taught us and the values they passed down to us. Our elders' words are deeply ingrained in our collective diasporic memories, as the grief of departing from our homes and Lands rarely fades away. It is through the shared pain of leaving home and the resilience we find on our journeys that common ground and strength emerge. Through this understanding, we can support one another in our ongoing struggle for recognition and justice. Our diasporic identity is therefore shaped by dislocation and longing for a homeland. Our identities are often marked by a blend of cultural heritage and adaptations to an unfamiliar environment. For example, in the diaspora, we might not have the same access to traditional foods, so we incorporate local ingredients into our meals. Tortillas, traditionally made from corn, have been adapted in various ways in the United States. We often adopt Western clothing, such as jeans and other garments, that we rarely wear in our countries. These are just some of the few examples of how we adapt and blend our cultures with our unfamiliar environments. We adapt to our present-day lives so that we can ensure that our youth are granted opportunities in these new Lands. Nonetheless, even in the diaspora, it is important to allow our youth access to our traditional norms and cultural values. Through our diasporic identity, we can create communities that celebrate and preserve our heritage. Creating an environment that embraces our roots ensures that future generations will remain connected to their heritage. This cultural continuity helps rebuild homes filled with Indigenous values, even while living far from our Indigenous Lands.

However, while we rebuild our homes, far away from our ancestral homelands, we may never feel truly welcome or wanted. Our sense of belonging in the new environment is often fragmented and incomplete compared to the deep-rooted connection we feel toward our homelands. In our ancestral Lands, every aspect of life is intertwined with our cultural traditions, shared history, and a profound sense of community. In contrast, the new environment lacks this intrinsic bond, leaving us in a perpetual state of yearning and displacement. Even organizations and places created for Indigenous folks can ring hollow when created by outsiders. In the diaspora, we are placed within Latinx spaces. Latinx spaces are environments where people of Latin American descent can come together to celebrate their culture, share experiences, and support one another. However, some Latinx spaces perpetuate false narratives about Latin American Indigenous culture or romanticize our identities, but only in the past tense. Our complex identities as displaced Indigenous Peoples are often oversimplified. In all these spaces we coexist in, so we must create our own spaces within these communities.

OaxaCalifornia is a great example of this, as it was a space created by pockets of community members and people representing different pueblos of Oaxaca. OaxaCalifornia serves as a vital cultural enclave where members can reconnect with their Indigenous Oaxacan heritage and traditions.[1] It provides a space for the community to engage in traditional practices, celebrate festivals, and pass down traditional knowledge to younger generations. This sense of cultural continuity and shared identity is crucial for maintaining their heritage and fostering a strong, supportive community amid diaspora challenges. It is critical that we not forget or ignore that, in what is now considered California, there are countless ancestral Lands of many Indigenous communities, Tribes, and pueblos. Gabrielino, Tongva, and Chumash, are some of the many Indigenous communities hidden in a state home to Hollywood glamour. Acknowledging these Indigenous ancestral Lands is not only a form of respect but also an imperative step toward recognizing the history and the past and current contributions of these Indigenous communities. It serves as a reminder of the deep-rooted connections to the Land that existed long before current state boundaries and modern developments. By honoring the presence and heritage of these Tribes, we foster a more inclusive and accurate understanding of the region's rich cultural tapestry.

We are forced to make homes in the diaspora, but these homes cannot ever be built on Indigenous erasure. My elders taught me to show respect not only to the Lands but also to the people whose ancestors stewarded those Lands before settlers stole them. As the adage goes, the United States is indeed a melting pot of immigrant cultures, but Indigenous Peoples continue to be displaced. In light of this, we need to recognize our shared responsibility in repairing the damage to Indigenous Peoples and their Lands while recognizing and honoring their Land rights. We must also protect Indigenous communities from further harm. This can include supporting Indigenous-led initiatives, advocating for policy changes that protect their Lands and rights, and educating ourselves and others about their histories and cultural practices. Building genuine partnerships with Indigenous communities based on mutual respect and collaboration is vital. Financial reparations and returning stolen Lands are crucial in acknowledging and rectifying past injustices. For those of us displaced Indigenous Peoples who are allotted privileges in this country, it is our responsibility to use those privileges to support our community members back home. This looks like sending resources, advocating for Indigenous rights, and amplifying Indigenous voices whenever possible.

It is through these actions that we can contribute to healing and justice for our communities. This is why I founded Earth Daughters (Se'e Ñu'un), an Indigenous-led collective that supports transnational Indigenous women and youth through mutual aid and climate justice initiatives. Earth Daughters focuses on providing necessary resources, fostering community solidarity, and advocating for systemic changes that honor Indigenous rights and traditions. By centering the voices and leadership of those most impacted, we aim to create a more equitable and sustainable future. Considering that not every community member can leave their ancestral Lands to seek out resources, Earth Daughters fosters a sense of collective support through mutual aid. We tirelessly lead fundraising efforts to help our community members when they need it the most. Our initiatives ensure that everyone has access to necessary resources and support, fostering resilience and solidarity. Through these efforts, we strengthen our community bonds and uphold our commitment to mutual aid. As displaced Indigenous Peoples, we cannot forget the layers of settler colonialism that exist in the United States as well as in our homelands. Recognizing these realities, our work not only provides immediate assistance but also challenges and addresses the systemic injustices faced by our communities. By doing so, we honor our ancestors and strive for a future where our rights and Lands are respected and protected.

Settler colonialism creates displacement cycles. Whether it is internal or external displacement, many of us are forced to uproot ourselves from the Lands our ancestors walked on. Climate change (rising temperatures, extreme weather events, and changing ecosystems) makes living off the Lands difficult and our traditional ways of living largely unsustainable. These environmental changes lead to food and water shortages that threaten the livelihoods of Indigenous communities. In response, Indigenous communities are employing traditional ecological knowledge and innovative practices to adapt to these ecological challenges.

They are engaging in sustainable agriculture, restoring native plant species, and advocating for policies that protect their Lands and resources. These efforts are not solely aimed at survival but also at preserving cultural heritage and strengthening community resilience. In many of our communities, the only method for climate adaptation is climate-induced displacement. This forced migration disrupts cultural ties and erodes the social fabric of our societies. Despite the efforts to adapt, the scale of environmental change often leaves us with no choice but to abandon our ancestral homes.

The immigration cycle that continues to dispossess Indigenous Peoples, especially those from the Global South, away from their ancestral Lands and kinships will always exist until we reach decolonization. This dispossession pushes them into the Global North countries (e.g., United States, Canada, etc.) that are responsible for backing and supporting atrocities committed against Indigenous communities and pueblos by settler governments and, on top of that, are constantly contributing to climate change impacts. Settler governments have historically enacted policies and practices aimed at eradicating Indigenous cultures and seizing their Lands for economic gain.[2] These governments often implemented forced relocations, assimilation programs, and violent suppression to diminish Indigenous Peoples' presence and rights. For instance, the United States government enacted the Indian Removal Act of 1830.[3] As a result, thousands of Native Americans were removed from their ancestral Lands, leading to the Trail of Tears. Residential schools in Canada were established with the goal of eradicating Indigenous cultures by removing children from their communities and subjecting them to harsh conditions.[4] Due to this systemic oppression, Indigenous communities have been stripped of their resources, cultural heritage, and autonomy, perpetuating cycles of poverty and displacement. Today, Indigenous communities continue to struggle with the repercussions of historical injustices and seek to reclaim their cultural rights. By neglecting sustainable practices and failing to address the root causes of environmental degradation, the Global North exacerbates the climate crisis, which disproportionately affects Indigenous Peoples. This environmental harm further undermines the livelihoods and traditional ways of life of Indigenous communities, who rely on the Land for sustenance. Addressing climate change is crucial in this context because Indigenous communities are often the most vulnerable to its impacts despite contributing the least to the problem.[5] Urgent action toward decolonization and ecological responsibility is required.

Global North countries continue to contribute to the increase in greenhouse gas emissions. This accelerates the climatic conditions that Mother Earth desperately wants to regulate but cannot. Yet, those in the Global South are the ones left to face the highest repercussions of the Global North's actions. An effective policy could be the implementation of stronger legal protections for Indigenous Land rights, ensuring that these Lands cannot be exploited without the consent of the Indigenous communities. Additionally, governments could establish environmental regulations that specifically address the unique vulnerabilities of Indigenous territories, promoting sustainable practices and penalizing activities that lead to environmental degradation. International cooperation and support for Indigenous-led conservation initiatives would further empower these communities to maintain their traditional ways of life while protecting their Lands from ecological harm. Still, the climate change discourse is filled with messages of doom or despair, overshadowing the resilience and wisdom of Indigenous communities. However, behind this discourse lies the responsibilities and accountability the Global North should take to truly mitigate climate change. The gloomy narrative often neglects the solutions and sustainable practices that have been part of our Indigenous traditions for generations. While the challenges are daunting, there is still hope if concerted efforts are made to adopt sustainable practices and support Indigenous-led conservation initiatives.

Many policymakers and researchers from the Global North remain skeptical of traditional practices, often viewing them as unscientific or outdated. The Global South's knowledge continues to be dismissed in discussions of climate change policies, regulations, and research despite our personal and ancestral experience with the natural world. Yet, within the climate justice discourse, our stories, our voices, and our faces are nonexistent. Indigenous knowledge offers valuable insights into sustainable Land management and biodiversity conservation, practices that have been honed over centuries.[6] This knowledge emphasizes a harmonious relationship with nature, promoting techniques such as agroforestry, controlled burns, and water stewardship. Integrating these practices into global climate strategies can lead to more effective and culturally respectful solutions to environmental challenges. However, the mainstream acceptance of Indigenous knowledge faces significant hurdles, including deeply ingrained biases and the marginalization of Indigenous voices. When Indigenous Peoples from the Global South are finally acknowledged, they become numbers, statistics, or percentages that are miscategorized. Overcoming these challenges requires a concerted effort to decolonize climate science and create spaces where Indigenous perspectives are not only heard but also valued and integrated into policymaking. As noted in my first book, *Fresh Banana Leaves: Healing Indigenous Lands Through Indigenous Science*, Indigenous knowledge systems offer the best climate change solutions. By embracing these time-tested traditional practices, we can address environmental challenges in a way that respects cultural heritage and fosters global cooperation. The future of climate justice hinges on the recognition and integration of these invaluable insights.

Global advocacy spaces are slowly recognizing Indigenous knowledge systems. Regardless, Indigenous Peoples, whether displaced or living back on their Lands, know that Indigenous science is already finding solutions to climate change. For instance, traditional fire management practices have been incorporated into modern wildfire prevention strategies in Australia. Additionally, Indigenous agricultural techniques are being used to enhance soil health and crop resilience in various parts of Africa. These examples highlight the valuable contributions of Indigenous knowledge to contemporary environmental and agricultural challenges.

One major question is: How do we uphold Indigenous science through displacement when Indigenous Peoples have been, and still are, forcefully disguised, ignored, or denied?

We can start by creating inclusive platforms where displaced Indigenous voices are heard and prioritized, as well as collaborating with Indigenous scholars and community leaders to document and share their stories of displacement.

1

Preparing the Soil:
Our Displacement

A s a young Indigenous girl, at the age of five, I began searching for where I came from. I looked so hard that I got lost. I felt a deep sense of longing and confusion as if I were wandering through a maze without a map. There were moments of frustration and sadness but also glimmers of hope whenever I uncovered a piece of my history. The journey has been as much about finding myself as it has been about discovering my roots in the diaspora. I knew who I was but not where I came from. This was confusing for a child to comprehend. I knew I was Indigenous, but I did not recognize my ancestral Lands as my home because I lived in a different place. The environment I now called home introduced new cultural norms and values that often clashed with my Indigenous heritage. For instance, my adopted home's emphasis on individualism and self-reliance conflicted with the communal and interdependent nature of my Indigenous community. Despite growing up in a new place, my family and community held our spiritual and nature-centric values close, so El Norte's fast-paced lifestyle and focus on material success felt alien. This made it difficult to reconcile the two, and I struggled to blend in while holding on to my traditions and stories. This duality forced me to navigate a complex landscape of belonging and self-acceptance. At such an early age, being labeled Latina or Hispanic prevented me from truly understanding how I, as an Indigenous girl from other ancestral Lands, could build an identity.

It wasn't until I fully grasped the notion of displacement that I could comprehend what the forced relocation of my roots meant. Until then is when I could finally understand where I truly came from. Learning my roots was challenging because visiting my relatives and families back in Oaxaca meant filling out immense paperwork and undergoing rigorous entry at the border. When I would visit Oaxaca, my parents had to get a letter detailing parental consent authorized by a public notary. This aligns with the life stories of many displaced Indigenous children whose parents sent them to visit their abuelitos. The obstacles that are in place to prevent the youth from never forgetting where they come from are intentionally constructed. This act of ensuring that children maintain a connection with their roots is a testament to the resilience and determination of our communities. It underscores the importance of preserving our heritage amid the challenges of living in a foreign Land. The Mexican consulate became a familiar place for my family. We often visited so I could travel back to my maternal home alone. It served as a crucial link between our lives in the diaspora and our ancestral Lands, as it was the place where we could get all the paperwork completed to facilitate my journey back to my maternal home. However, what happens to children who bear the brunt of being undocumented in this country? They face additional layers of fear and uncertainty, constantly worrying about separation from their families while not being able to visit their Lands. I must acknowledge the privileges granted to me for being able to go back to my ancestral Lands and grow up in close connection with them. Undocumented children do not have this privilege. This stark contrast in experiences highlights the disparities within our communities and the need for compassionate immigration policies.

As I've grown older, I've come to realize the importance of guiding our displaced Indigenous youth. We must become mentors, helping them navigate their identities and maintain a connection to their heritage. By sharing our experiences and wisdom, we can ensure they feel rooted and supported despite the challenges they face.

When I was back in Los Angeles, California, I was desperate to return to my true home, to feel their familiar embrace. So I can only imagine what children who are never allowed to step back into their ancestral Lands feel. I correlate our sense of belonging to our emotional and mental health. It is crucial to address these disparities and work toward creating a world where everyone can feel connected to their roots and heritage.

One way communities can support displaced Indigenous youth is by creating cultural programs that celebrate and teach them our traditional practices. Establishing community centers where families can gather, share stories, and find support can also nourish a sense of belonging and emotional well-being. Creating safe spaces for our youth is essential. These spaces can offer sanctuary where young people can express themselves freely and connect with others who share similar experiences. By fostering such environments, we can help our youth build a stronger sense of identity and community.

As a young girl, I felt alone in a world that labeled me Latinx or Hispanic or migrant. It took me awhile to realize how harmful these labels were to my identity as a displaced Indigenous person. Throughout my experience, I felt anxiety and depression. It was only after visiting my maternal Lands that I understood how our Lands support our emotional and spiritual well-being. As soon as I understood my displacement, I rebuilt a sense of connection with myself. It took me years of reflection and learning to grasp the complexities and nuances of what displacement meant to my overall well-being. It is for this reason that we should support one another in telling our stories as Indigenous Peoples who have been displaced. Our youth need these stories, so they do not feel alone and can start to realize that what they are dealing with is not a unique experience.

Instead, it is an experience we all share as displaced Indigenous Peoples. By sharing our stories, we can provide our youth with a sense of solidarity, resilience, and hope for the future. The role our communities play in overcoming displacement cannot be overstated. It is within the collective embrace of our communities that we can find strength, understanding, and a shared sense of purpose, and together, we can create a support system that empowers each of us in reclaiming our heritage and healing from our traumas of displacement.

As Indigenous Peoples, we view our Lands differently, which explains why I felt this sense of loss and lack of connection in a foreign Land. Being stripped of our Lands is more than just physical displacement; it disrupts spiritual and cultural ties, leaving us uprooted and disconnected. This profound sense of grief permeates our communities, impacting our mental and emotional well-being. It often manifests itself as depression or anxiety, as individuals struggle with identity fractures caused by displacement. This disruption can lead to loss of belonging and instability, further exacerbating emotional turmoil. Our loss of identity, heritage, and our sacred relationship with the earth is something that deeply impacts us.

While I did not understand what my spiritual and emotional being needed when I was young, my parents did. They made sure I could visit our ancestral homelands even when they couldn't. They instilled in me the importance of our traditions and ancestors' stories, ensuring a strong connection to our roots. My parents saved funds so that we could visit our ancestral homelands, allowing my siblings and myself to experience the sacred spaces that held and continue to hold our histories and cultures firsthand. Those kinds of efforts preserve our identity and pass down the knowledge and values that define us as a people. In doing so, my family also hinted at the sacrifices they made, and continue to make, for us to maintain our Indigenous cultures and identities. Each visit felt like a homecoming, filling me with the sense of belonging and peace I had been longing for. These experiences had a profound effect on me, not only through reconnecting me with our Lands, deepening my connection with the wisdom and resilience of our ancestors, but by fueling my commitment to preserving our heritage for future generations.

Sometimes the sacrifices of our elders, families, and communities remain silent, especially concerning immigration status. Their status is meant to be hidden and kept as a family secret. We are told not to trust anyone with this information. This creates an environment of fear and isolation, where even close friendships can be tinged with anxiety. The weight of this secrecy often shapes a child's understanding of loyalty and caution.

Throughout my childhood, I was told to never speak of my parents' status, especially my mother's. I was in a constant sense of vigilance, always careful about what I shared and with whom. It was a burden that made me grow up fast; I understood the delicate balance between trust and protection. I did not comprehend what nationalities meant, and I did not grasp why my mother, as a Mexican national, had to await a longer immigration process than my father, who was a Salvadoran national. My dad was granted asylum due to the heavy involvement of the United States in the Central American Civil War.[1] This disparity between my father's and mother's immigration process puzzled me as a child and added to our family's intricate tapestry. It was only later that I understood the intricate geopolitical factors that influenced our lives so profoundly. This is why I am called to acknowledge the privilege I had growing up with access to my maternal Lands. Recognizing this privilege has deepened my commitment to advocating for displaced Indigenous Peoples' rights and cherishing and protecting our cultural heritage. For my ancestors, my family, my parents, and the young children who also face this sense of disconnection, the ancestral Lands they were forced to leave.

My mother, Juana, an Indigenous Oaxacan woman, was unable to see her mother for over twelve years. She had to wait years for an update on her immigration status to see her mother again. Can you imagine not seeing the woman who gave birth to you for that long? Sadly, many Indigenous Peoples can. Her story of separation and reunion reminds us of the importance of preserving our connections to family and heritage. Despite this, she carried my abuelita's teachings deep in her heart. Even when we were far from abuelita, my mother made sure we never forgot who we were and where we came from by raising us with her mother's wisdom and traditions, instilling a strong sense of identity and cultural pride. These teachings shaped my values and guided my actions, fostering a deep respect for my heritage. They inspired me to learn my ancestors' Indigenous languages and customs, ensuring our family's rich cultural tapestry continues. Moreover, they instilled a responsibility to pass these traditions on to future generations in me. This dedication to preserving culture is common among Indigenous parents. They strive to keep their traditions alive, even when faced with displacement, ensuring their heritage endures despite the challenges. Resilience and commitment are vital for their cultural identity.

For displaced Indigenous Peoples, not remembering the winds that whispered songs to them while they were children playing in nature is a profound loss. The sounds of native birds that once greeted them every morning have become distant memories, creating a longing that never leaves them. The separation often results in lost connections and a deep sense of longing for their cultural roots. Unfortunately, this longing is not considered in their immigration cases for residency, citizenship, or asylum. This complicates their journey to a country that does not, and may never, feel like home. Twelve years after leaving her ancestral Land, my mom reunited with her relatives. During those twelve years, my mother faced numerous legal challenges, including navigating a complex and often unforgiving immigration system. She had to deal with lengthy application processes, frequent changes in immigration laws, and the constant fear of deportation. The high cost of legal representation and the emotional toll of prolonged separation from her family made the journey even more arduous. Phone calls become the only connection Indigenous Peoples have with their relatives back home, serving as lifelines to their present. Videos become their second pair of eyes that allow them to witness their

Lands transform through the years, bridging physical distance with digital memories. These virtual connections, while invaluable, are reminders of the life and Lands they are separated from. Though my mother never lost contact with them, virtual connections could not replace the physical hugs, embraces, and love she longed for. That reunion was the bittersweet culmination of years of separation. It was a testament to her resilience and the strength of family bonds. As soon as my mother saw her sister and her beautiful mother in person, no one could separate them. They laughed and reminisced about countless summer days they spent together, family gatherings filled with joy, and late-night conversations that brought them closer. Their shared bond was evident in their eyes, filled with love and nostalgia. Every time I returned home to my maternal Lands, my grandmother would tell me how she saw my mom in me. She also shared that she looked forward to seeing her daughter again. As she spoke her eyes sparkled with longing and hope, and I could feel her love for both of us. My mother's reunion with her mother, my grandmother, was filled with a deep sense of connection and belonging. Witnessing the love and joy between my mother reminded me of the importance of cherishing these precious moments and the legacy of love that binds us together; their connection shaped me. Like papayas, my mother's reunion with her mother reminded me that while it was a journey filled with a mixed array of emotions, it was something irreplaceable. The bittersweet moments, the laughter, and the tears all came together to create memories that would last a lifetime. Just as the unique taste of papaya cannot be replicated, neither can the deep-rooted bond of family.

As a result of my mother's displacement to this country, the United States, she had to sacrifice a lot. When my grandfather joined our ancestors, my mother did not get to say farewell to him. I was five years old and had chicken pox, so she did not want to leave my brother and me behind while she traveled back home, although she was willing to risk taking the dangerous journey back to the United States. Missing the chance to say goodbye to her father left a void in her heart that could never be filled. This unspoken farewell became a lingering sorrow, affecting the strength of family ties and deepening the emotional scars of separation. The inability to share such crucial life moments underscores the profound sacrifices made by migrant families, revealing the true cost of their pursuit of a better life. My mother cried for days, wishing she could participate in his funeral. However, her maternal love would not allow her to leave us while we were ill. Her sacrifice deepened her longing and sorrow, adding another layer to the complex emotions faced by immigrant families separated by borders and legalities. When I was young, I did not understand why my mother was covered in tears every day. Still, at my tender age, I knew she was grieving; her sorrow was palpable. It was a poignant lesson in the heartache of the immigrant experience and an example of the hard decisions and sacrifices Indigenous Peoples make when they leave their homes. It was a powerful lesson that I will never forget. As I grew older, the memory of my mother's grief became a touchstone for my own understanding of sacrifice and resilience, taught me the importance of family, and instilled in me a profound sense of gratitude and a relentless drive to honor the sacrifices of those who came before me.

These are the stories we are taught to keep to ourselves and rarely get to read, as beneath these stories lie the immense grief our families still endure. By sharing these narratives, we honor the strength and spirit of our ancestors and acknowledge the ongoing struggles displaced Indigenous communities face. I have experienced the impact of displacement on Indigenous Peoples through the loss of my uncle, Leonardo Betanzos Santos. He left our beautiful Oaxaca because he was having a hard time finding a job and wanted to support my grandmother after my grandfather passed away. Despite his best intentions, the distance created a chasm that has been difficult to bridge, leaving him with an enduring ache for the home and family he left behind. My parents supported him on his journey to the United States and I grew up with him. I saw him as an older brother as he was younger than my mother. The second youngest of nine children, he appreciated his mother, my grandmother. He wanted to ensure she had a comfortable life as an elder. He always felt a sense of responsibility, given that my grandmother was left to raise him and my youngest uncle when my grandfather passed on. This profound sense of duty shaped his decisions and sacrifices, driving him to seek better opportunities far from home. Yet, the weight of this responsibility also meant living with the pain of separation, a burden he bore with quiet strength and unwavering resolve. My uncle never married, and unfortunately, when my grandmother passed on, he was unable to say goodbye to her in person. Despite knowing that my grandmother's transition was a part of life, the inability to be there during her final moments added another layer of sorrow to his already heavy heart.

This experience serves as a poignant reminder of the sacrifices and emotional toll that displacement carries for many Indigenous families. My grandmother gave her goodbye to my uncle over the phone, and the call was met with a heart-wrenching silence on my uncle's end of the line. He was deeply grieving and longed to see the mother who had raised him single-handedly after my grandfather's passing. This final farewell over the phone intensified his sorrow, highlighting the emotional toll that displacement carries for many Indigenous families.

Two years after my grandmother's passing, my uncle had a fatal heart attack as he was getting ready to visit us for the holidays. The news of his sudden death was devastating, and his absence was deeply felt; it left our family with no desire to celebrate. The festive season, once filled with joy and anticipation, turned into a period of mourning and reflection. Our planned posada became a funeral, but my uncle did get his last wish: to return to his Lands. In a solemn ceremony, we laid him to rest in the place he always considered home, surrounded by the family and community he loved so deeply. This final act of bringing him back to his roots allowed us to honor his memory and the sacrifices he made for our family. The story of my uncle resonates with many displaced Indigenous Peoples who ask their loved ones to return them to their Lands when they leave this world.

Many Indigenous Peoples are unable to return home in their physical form, but their remains, ashes, and spirits can.

As Indigenous Peoples from the Global South, we are not here to gatekeep Indigeneity or our Indigenous cultures. Instead, we are here to remind everyone that Indigeneity isn't merely the romanticized version of our culture many discourses advocate for. We strive to foster a deeper understanding and appreciation of Indigenous complexities and realities. We seek to share our stories and struggles to ensure our cultural heritage is respected and preserved. Our goal is to bridge gaps and build solidarity. We recognize that Indigeneity encompasses a wide range of experiences and histories that are not imagined or simply points in history. It is about the real lived experiences of sacrifice, separation, and resilience. All of our stories, like my uncle's story, reflect the profound challenges and enduring connections to our Lands and communities, especially through displacement. Reconnecting with our roots is a valid and essential process, especially for our Indigenous youth who have faced forced displacement through removal from their cultures due to adoption and other disruptions. We understand the importance of this journey and support their efforts to reclaim their heritage and identity. Through acknowledging these struggles, we aim to create spaces where healing and reconnection are encouraged and facilitated. However, when these narratives are used to amplify someone's career or personal gain, it can leave a lasting impression on those who know all too well what forced displacement looks like. It trivializes our experiences and undermines the genuine hardships we've faced. Our stories are not mere stepping-stones for others but sacred testimonies of our resilience and connection to our heritage. Cultural appropriation further complicates these struggles by commodifying and misrepresenting our cultures. When elements of our traditions are taken out of context and used for profit, it is not only disrespectful to our heritage and traditions but erases the true meaning and significance behind them. This exploitation can cause deep emotional pain and perpetuate the very disconnection we are striving to heal.

With age, we understand the sacrifices our parents, family members, and communities made for us. For those of us who had to leave our Lands at a young age, we understood that our displacement was the only way we could survive. This realization brings a deep appreciation for their courage and the difficult decisions they had to make to ensure our well-being and future. Coming to terms with displacement is an emotional journey filled with gratitude, guilt, and understanding. As we reflect on our parents' sacrifices, we grapple with cultural loss and disconnection from our roots. We also grapple with the identity shifts from growing up in an unfamiliar Land. This journey is marked by moments of profound empathy and respect for the resilience that has shaped our lives. This leads us to cherish the heritage and legacy they preserved through their hardships. They inspire a commitment to preserving cultural traditions and values, ensuring they are passed down to the next generation. Moreover, they foster a deeper connection to one's roots, bridging the gap between the past and the future.

These teachings and understandings are reasons why the younger generation is more vocal about not accepting these forced identity labels. They strive to create a space where Indigenous identities coexist without imposed definitions. This defiance is a tribute to resilience learned from their ancestors, championing a future where cultural heritage is celebrated in its full complexity. We also find the courage to decline identity labels given to us by Western countries, such as the United States. Our experiences are lumped together under the umbrella of a homogeneous Latinx experience, but as Indigenous Peoples, we have our own narratives. Our deep connection to our Lands is more profound than what that broad label can embody. Indigenous identities are deeply rooted in specific Lands, traditions, and cultural practices that have existed for millennia. Unlike the broader Latinx experiences, which encompass a diverse range of cultures and histories across Latin America, Indigenous identities are tied to unique ancestral knowledge and spiritual connections to their territories. This distinction emphasizes the importance of acknowledging and respecting the individuality and sovereignty of various Indigenous Peoples. Our identity is rooted in a heritage that transcends labels.

Navigating cultural assimilation in another country often means balancing the preservation of our Indigenous heritage with pressures to conform to mainstream norms. Despite these challenges, many find strength in our traditions and community. We use them as a beacon to guide us through the complexities of maintaining our cultural identity in a foreign Land. For example, traditional ceremonies and rituals are often adapted to fit our new urban settings, such as Xandú (Day of the Dead). In urban areas, where we may be far from the resting places of our ancestors, we create altares in community centers or in our homes to honor them. This adaptation allows us to keep the spirit of Xandú alive while embracing our new environments. This allows our community members to gather and celebrate our heritage in city parks or community centers. Indigenous cuisine is shared and enjoyed at cultural festivals, introducing traditional dishes to new audiences while preserving culinary practices. Language preservation efforts, such as community language classes and online resources, help younger generations maintain their native tongues despite being far from their ancestral Lands. Understanding of Indigeneity pulls us into these spaces. Still, we must walk within these spaces respectfully. The growing availability of cultural spaces and resources fosters a sense of belonging and pride among our people. This growth gives hope to all Indigenous communities in the diaspora.

In Latinx spaces, we are often invisible, as Indigenous Peoples are frequently discussed in the past tense. This invisibility stems from colonial histories that sought to erase Indigenous identities and cultures, favoring a homogenized national identity. The legacy of colonization continues to marginalize Indigenous voices, leading to a lack of representation and recognition within broader Latinx narratives.[2] By reclaiming our stories and advocating for visibility, we challenge these historical erasures and assert our rightful place within the diverse tapestry of Latin American cultures. As Indigenous Peoples, if not ignored, we are romanticized, only seen as parts of history that nationalism exploits for commodities such as tourism. Yet, our present and future are rarely acknowledged within Latinx spaces. This erasure perpetuates stereotypes and ignores Indigenous communities' ongoing contributions and struggles. Consequently, we are compelled to carve out our own narratives and advocate for recognition and respect for our living cultures. This is why our Native relatives from the Global North embrace us, as our histories and present-day lives parallel one another. The only things that separate us are the settler borders that cross our Indigenous Lands. By uniting with our relatives across these artificial divides, we amplify our voices and strengthen our resilience in the face of ongoing colonization.

My memories of displacement are perfumed by the aroma of papayas. When I helped my mom peel and slice papayas, she often shared cherished memories of her childhood in her homeland with me. These stories intertwined with the smell, creating a poignant connection between the fruit and our family's past. Each slice carried the weight of her nostalgia and longing. The smooth, waxy skin of the papaya would yield under the blade, revealing the vibrant orange flesh beneath. As I cut through the fruit, the sweet, musky scent would fill the air, mingling with the distant echoes of my mother's stories. The seeds, slippery and black, would tumble out, adding a tactile element to the sensory experience, making each moment a poignant reminder of our heritage and the bittersweet journey of displacement. The scent evoked a blend of warmth and sorrow, filling me with a feeling of belonging among a yearning for a place I had never known. It stirred empathy for my mother's, for our, displaced lives. The aroma was both comforting and haunting, an olfactory bridge to a distant, almost mythical past. During our moments together peeling and slicing papayas, my mother's eyes often glistened with joy and sadness. She found solace in sharing her stories, yet there was an underlying pain of longing for a home that was no longer hers. Her voice wavered slightly, revealing her deep emotional ties to her homeland. People's sentiments around papayas are influenced by their life experiences and cultural backgrounds. For some, the fruit symbolizes tropical paradise and indulgence, while for others, like my mother, it carries a more profound significance. A simple fruit can evoke diverse and powerful feelings.

Like papayas, my mother always told me, displaced Indigenous Peoples were either loved or hated. This stark, binary reaction mirrored the complex emotions tied to our heritage and immigration impact. Papayas serve as a reminder that our identity is often met with extremes, much like the fruit's divisive taste. Papayas, like immigration, evoke strong reactions with little neutrality. People are taught to develop firm preferences, especially regarding food. This extends to their views on cultural integration. This polarization can lead to significant societal divisions, affecting the sense of unity and cohesion within communities. When people take extreme stances on immigration, it often results in policies and attitudes that either strongly support or vehemently oppose the integration of immigrants, leaving little room for nuanced understanding. This lack of acknowledgment of the complexities of immigration can lead to policies that are ineffective or even harmful to immigrants and their communities. Such divisive perspectives can hinder constructive dialogue, perpetuate stereotypes, and create barriers to the mutual acceptance of diverse cultures.

In our family, papayas were more than just some fruit; they symbolized resilience and connection to our roots. My mother often recounted how her own parents cultivated papayas as a staple in their diet, a link to their ancestral practices. She would explain how they carefully selected the seeds, planted them in nutrient-rich soil, and ensured they received plenty of sunlight and water. The trees were tended to with great care, pruning them to encourage healthy growth and protect them from pests. Harvesting the ripe, golden papayas was a family affair, marking not just the fruition of their hard work but also the continuation of a cherished tradition. The act of peeling and slicing papayas became a ritual that bridged generations, preserving our cultural heritage despite displacement fractures. In our new Land, we could not grow the same papayas that surrounded my mother's home, but we could slice them together, savoring the sweet taste and reminiscing about the stories tied to each piece. This simple act brought us closer, allowing us to honor our past while creating new memories in our present. Even in a different Land, the essence of our heritage remained alive through these small but meaningful traditions.

However, during my teenage years, the once comforting aroma began to remind me of feet, and I became increasingly repulsed by the taste. After this shift in perception, I questioned why I felt this way. I came to realize it was a subconscious act of rebellion. I was not mad at the papayas themselves; I was mad at the world. I was angry that my family was displaced from our Lands, and I felt like I had no choice in the matter. Rejecting the fruit seemed like a small way to assert control over my identity and experiences, even if it was just a symbolic gesture. My family, although initially puzzled by my aversion, accepted it with understanding and patience. They realized that my rejection of the fruit was not about taste but about grappling with the broader struggles of our displacement. Through my aversion to papayas, my family came to recognize the personal battles I was facing in adapting to a foreign environment. This mutual understanding fostered a deeper empathy and connection among us, strengthening our familial bonds despite the challenges we encountered. The papaya became the symbol of my displacement, embodying the complexities of my uprooted childhood. This duality made every encounter with the fruit a loaded experience, as it embodied both the loss of our homeland and the enduring strength of our family ties. Eventually, I came to appreciate this complex relationship, realizing that embracing the papayas symbolized accepting our shared experiences as beings.

During a visit to my maternal homeland, I was unexpectedly drawn to a fresh, ripe papaya at a local market. When I bit into the fruit, the familiar taste evoked memories and emotions. I finally understood my mother's deep connection to papayas. At that moment, I began to appreciate fruit not just as food but as a symbol of our resilience and enduring heritage. Being able to peel and slice papayas alongside my grandmother showed me why my mother enjoyed doing this with me. It was the maternal love she felt and could share through this ritual, a way to pass down traditions and stories through simple, shared actions. I cherish the moments we spent peeling and slicing papayas together.

By sharing stories, even ones as simple as enjoying a fruit together, we reclaim our voices and promise that our rich cultural tapestry is preserved. These are similar tapestries to our traditional clothes, such as our huipiles. While they are filled with sorrow, longing, and grief, our resistance and resilience fill them with colors. Each thread woven into our stories adds a vibrant hue, symbolizing our strength and the unbreakable bonds that connect us to our ancestors and culture. This act of storytelling is a powerful form of resistance, asserting our identity and presence despite the miles that separate us from our ancestral homes.

Markets in my maternal pueblo thrive with vibrant colors and the lively chatter of vendors and customers. You can hear the chisme shared throughout space. Stories of neighbors, family updates, and local happenings flow freely, creating a tapestry of community life. This tradition makes shopping interactive and fosters community and mutual respect. Community bonds mean we catch up on family stories and updates. The market stalls are always brimming with a dazzling array of fruits, each more enticing than the last. Juicy mangoes, plump with sweetness, sit beside fragrant guavas and tangy tamarinds. Vibrant oranges, luscious pineapples, and creamy avocados add to the colorful display, inviting shoppers to savor the rich bounty of the Land. It reflects the deep-rooted customs of the people and the importance they place on personal connections even in everyday transactions. The air is always filled with the rich aroma of spices and tropical fruits, creating a sensory tapestry that reflects our community's heart and soul. Rarely do artificial smells fill the atmosphere, as our traditional foods and diets are abundant in nutritious, whole foods. This dedication to natural ingredients and time-honored recipes nourishes our bodies but also preserves our cultural heritage. Each market visit celebrates our roots, a reminder of our values and flavors.

This bustling atmosphere is more than commerce alone; it is a celebration of our culture, a place where stories and traditions are exchanged and preserved. The humidity, which causes our pores to drip with sweat, does not seem so intense when we remember it is like the heat that stuck to our ancestors' skin in the natural saunas our ancestors used for spiritual cleansing for centuries. As we experience our pueblo markets, we sweat, but our sweat cleanses our souls and our spirits. Unfortunately, our pueblo markets are missing in the diaspora. While our communities have built adaptations of our Indigenous pueblos, there is always something missing.

One thing that is not different is that these adaptations of our pueblo markets have brought our traditional dishes with them so our sazón will be inherited by Indigenous Peoples for generations. Our unique flavors and culinary traditions continue to thrive, even after displacement. Our sazón adds a distinct blend of flavors and spices integral to our traditional cuisine. It enhances taste and aroma, making each dish a true representation of our cultural heritage. Without it, food lacks depth and authenticity. However, authentic flavors and traditional cooking methods that hallmark our cuisine are difficult to replicate outside our homeland. This is especially true when there are so many restrictions on food. The food regulations in the United States often force us to make and sell overprocessed foods in the name of sanitation and compliance. This compromises the authenticity of our dishes, making it challenging to preserve the flavors and techniques that contribute to our culinary heritage. These policies and regulations not only dilute the genuine flavors of our traditional dishes but also disconnect us from our ancestral practices passed down through generations. The meticulous techniques and ingredients that our ancestors used are often replaced with more standardized and less flavorful alternatives. This affects our culinary identity and erodes a vital link to our cultural heritage.

This can lead to our recreated markets in the diaspora sometimes feeling like shadows of what they once were. For example, our authentic mole recipe cannot be recreated with the Doña Maria canned version. This pre-packaged version lacks the complexity and richness of homemade mole. Mole is traditionally handmade with dried, thick-skinned chilhuacle chiles and ancho chiles toasted on a comal. This intricate process involves grinding the chiles with a variety of spices and ingredients, resulting in a rich and complex sauce that is central to many of our celebratory meals.[3] The deep, layered flavors of the homemade mole are a testament to the skill and dedication required to honor our culinary traditions. Nothing compares to the experience of enjoying freshly made molito after hours of toasting and grinding chilhuacle and ancho chiles along with a medley of spices at our pueblos markets. True mole preparation requires a meticulous process of blending diverse ingredients to achieve a harmonious balance. Our mole's essence cannot be replicated by mass-produced alternatives. The vibrant atmosphere and the sense of community that comes from preparing and sharing these traditional dishes are irreplaceable. This communal effort and the resulting flavors are a true reflection of our rich cultural heritage. In our celebrations, the mole holds a place of honor, representing unity and tradition. In festivals, weddings, and family gatherings, its preparation becomes a communal activity that brings people together. Our shared history and cultural identity are also highlighted by the rich and complex flavors of mole. In the diaspora, mole becomes a way to continue our celebrations while honoring our roots that have extended this far.

It is no surprise that the carritos that sell these traditional foods are often vandalized and discarded by policy enforcement. These small vendors represent a direct link to our cultural heritage and the flavors that define it. If we lose these carritos, we risk losing an essential part of our culinary identity and the communal spirit that comes with it. Street vendors are illegal in the United States; this makes it impossible to truly replicate the vibrant atmosphere of our pueblos markets. Despite the challenges, these carritos provide us with the authentic flavors and communal spirit central to our culinary heritage, even in the face of legal obstacles. The authorities seize the grills and carts, and it can be difficult for our undocumented relatives to pay the fines and fees required to get them back since these processes often require identification. This punishment not only affects their livelihoods but also hinders their ability to share our cherished culinary traditions with the community.

I remember, as a little girl, that everyone on the block knew they could come to ask my father to recuperate their grills or carritos. Because my father had been granted asylum in this country, he had the necessary identification and resources to navigate the legal processes. His efforts helped members of our community sustain their livelihoods but also ensured that our rich culinary traditions continued to thrive. At the time, I did not fully understand why my dad was asked for such favors, but he would always drive his troquita with our community members to retrieve the carritos and grills. In Los Angeles, getting a permit to be a street vendor was impossible due to the hefty fee.[4] Many aspiring vendors were met with barriers, making it difficult to start their businesses. Even as a child, I helped community members fill out these applications. I saw the struggles they dealt with firsthand and felt a deep sense of responsibility to assist them. Many Indigenous children have similar experiences; we become our parents' translators and often fill out rigorous paperwork way beyond our years. This responsibility can feel like a heavy burden on our shoulders, as if my community's future was dependent on our ability to navigate complex systems through these documents. We gain a sense of maturity and urgency, along with pride in helping with adult responsibilities, but at the cost of our childhoods. At the time, we did not have access to the internet; we could only rely on our elementary school education to get us through these tasks. We had to pull out our huge dictionaries to understand the bureaucratic language and long, foreign words that filled these official documents. We did not know how to answer what some of these application questions asked for. Each form we filled out was a microcosm of the obstacles we had to overcome to survive in the diaspora.

While community members often saved enough money to get these permits, the Department of Public Health (DPH) refused to approve their carts, and there was a huge lack of transparency coming from them.[5] Since most county health codes were written for physical establishments such as restaurants, they are designed for these types of establishments. Traditional establishments often face stringent regulations and requirements for food preparation and sales. As a result, many of our community members faced significant challenges in sharing the fruits of their labor with the broader public. Still, most vendors knew they could rely on my dad, and his troquita, to retrieve their carts and grills when needed. His dedication to helping our community overcome these obstacles was a testament to his perseverance and resourcefulness. This sense of solidarity and mutual support made a significant difference in our ability to thrive. While Los Angeles has created more accessible permit policies and reduced the permit fee, it is a part of our history that cannot be forgotten or ignored. These changes, although beneficial, do not erase the struggles and resilience that once defined our community's journey. Remembering these experiences is crucial in honoring our past and understanding the progress we've made.

I witnessed Indigenous solidarity at a young age. Now, we are asked how to build Indigenous solidarity. The answer is simple. It lies in our everyday actions and commitments to support one another despite where we come from. It is a testament to where our intentions lie, especially in the diaspora. My dad did not primarily work with Central American, Indigenous street vendors, but many were from Mexico and South America. It is interesting how we homogenize our experiences in the diaspora. Logically, we understand that our experiences are completely different and that we have different values and traditions, but we still simplify ourselves. Yes, there is xenophobia even in our communities, especially when it comes to Indigenous Peoples from other nationalities, but I was taught that these were narratives made to create separation and pit us against one another. At the end of the day, our diaspora narratives parallel each other. In times of despair, we truly understood what Indigenous solidarity means.

Indigenous solidarity is crucial in supporting and uplifting displaced Indigenous Peoples' rights and well-being. Yes, we come from different Lands governed by different settler colonial structures, but our histories and present remain linked. Indigenous Peoples in Mexico face significant mistreatment and marginalization, despite the pervasive myth that everyone in the country has Indigenous ancestry. This false narrative often serves to erase the unique struggles and identities of Indigenous communities, further entrenching systemic inequalities. Solidarity is vital to challenge these injustices and honor the true diversity and resilience of Indigenous Peoples. Unfortunately, these false narratives dominate what is known and believed in the United States. It is important for us to collectively listen to the stories of Indigenous Peoples from the Global South, both displaced and those still living in their Lands. By amplifying their voices and experiences, we can dismantle these misconceptions and work toward true solidarity and justice. When Indigenous Peoples are displaced from their ancestral Lands due to various reasons, such as forced displacement, development projects, or armed conflicts, they face unique challenges that require collective effort to overcome. For instance, grassroots organizations in the United States have created support networks that provide legal aid, housing, and cultural preservation programs for displaced Indigenous communities. Similarly, in Oaxaca, Indigenous pueblos have created collective efforts to help displaced Indigenous Peoples from Central America. These communities have established support systems that offer shelter, food, and legal assistance to those in need. By working together, they not only provide immediate aid, but foster a sense of unity and resilience among Indigenous Peoples across borders. Indigenous solidarity involves recognizing, supporting, and advocating for their rights, Lands, and resources, and promoting cultural preservation and revitalization.

We can start fostering solidarity by recognizing the challenges that displaced Indigenous Peoples face. These challenges include loss of cultural identity, limited access to resources, discrimination, and inadequate housing, healthcare, and education. Indigenous Peoples who have been forcibly relocated from their ancestral Lands often face difficulties maintaining their traditional practices, customs, and languages. Individuals and organizations must acknowledge the hardships experienced by displaced Indigenous Peoples and provide support accordingly. Even in the diaspora, displaced Indigenous Peoples continue to advocate for the protection of their ancestral Lands. This ongoing commitment highlights the deep connection to our heritage and the importance of preserving our cultural and environmental legacies. Through international collaborations and activism, they strive to have their voices heard and their rights upheld globally. Protecting Indigenous Lands and resources is a fundamental aspect of Indigenous solidarity. Indigenous Peoples often face threats from development projects, Land grabs, and environmental degradation.[6] One prevention strategy may involve advocating for legal protections, such as Indigenous Land rights, and opposing projects that pose risks to Indigenous communities. Indigenous solidarity means supporting education and cultural heritage efforts for displaced Indigenous Peoples. Indigenous knowledge and traditional practices hold immense value and should be preserved and passed on to future generations. By investing in educational programs, cultural preservation initiatives, and language revitalization programs, we can ensure that displaced Indigenous Peoples continue to maintain their cultural identity and knowledge systems.

This is the beauty of the upbringing many of us displaced Indigenous Peoples have, that nationality and boundaries do not define us. We are united in our struggles and shared history. My mother emphasized that despite the hardships and barriers, we must continue to support each other and uphold our shared values. My mother constantly reminded me that, like papayas whose journey through export and import is treacherous, all displaced Indigenous Peoples have made a similar journey. Papayas are typically imported through a complex supply chain. This involves harvesting the fruit at optimal ripeness, packaging them carefully to prevent damage, and shipping them via air or sea freight.[7] They often pass through multiple checkpoints and inspections to ensure they meet safety and quality standards before reaching their destination in grocery stores. This journey can sometimes compromise the quality of the fruit. This leads to an influx of less desirable papayas in grocery stores most accessible to Indigenous communities. The best papayas are often reserved for stores catering to wealthier demographics. This disparity not only affects the consumer experience but also highlights the inequalities inherent in our food distribution systems. It's no surprise that the smells emanating from such high-scale grocery stores overshadow the ones we usually don't find in the grocery stores designed for our communities. However, our communities resist these inequities through actions, through doing things our way, communally. This is why carritos have played a vital role in providing us with tasty traditional dishes or fresh fruit, like what we have access to in our pueblos markets. These mobile vendors not only bring high-quality produce directly to us but also preserve our cultural heritage and ensure that everyone can enjoy the authentic flavors of our homeland. Unlike papayas, no Indigenous person is less worthy, because of the arduous journey they took; as stewards of the Lands, we deserve the best fruit, the papayas of our ancestors.

When it comes to eating store-bought papayas, yes, the flavor may be there, but the sense of connection is missing. It is not the same as nurturing the trees that grow papayas and allowing us to connect with the fruit as it grows. The experience of tending to the trees and witnessing the fruit's development creates a deeper appreciation and stronger bond with the food we consume. This connection adds a layer of fulfillment that extends beyond mere taste. Papayas are native to the tropical regions of the Americas, particularly southern Mexico and Central America. Indigenous Peoples have cultivated and enjoyed these areas for ages. I remember summer days spent in my grandmother's backyard, where she had a small but flourishing papaya tree. We would eagerly wait for the fruits to ripen, and she would teach us to pick them carefully. The fragrant smell and the sweet taste of freshly harvested papayas always bring me back to cherished moments of family bonding and intergenerational learning. Understanding how papayas preserve memories of my grandmother and her mother, I can see how they are a looking glass through which we can relive maternal care.

Understanding how we, as Indigenous Peoples, share similar origins adds another layer of respect and appreciation for this remarkable fruit. As a result, when we see papayas at our tables, we are reminded of our Lands and our homes, where our true hearts lie.

2

Uprooting Our Roots:
Climate Change

Uprooting our roots as Indigenous Peoples does not mean we lose our identities or are no longer Indigenous. Being Indigenous means being aware of our communities, our Lands, and the traditions that keep our cultures alive. In Latin America, our identity is deeply rooted in our family and our close ties, rather than formal Tribal enrollment. A strong sense of belonging and connection to our ancestry helps preserve our cultural heritage and reinforces our Indigenous identity. These familial and communal bonds ensure that our traditions and values continue to thrive, regardless of where we reside. This is the beauty of being Indigenous. We are not just beautiful people with colorful traditional clothing, intricate beading patterns, or vibrant dances. We are people who have a strong foundation. Our roots are deeply embedded in our existence, which fuels our ongoing resistance. Despite everything we have undergone and continue to undergo, our roots will always be the foundation of our Indigeneities, and our roots will never be destroyed. Our connection to our heritage and the resilience of our spirits ensure that we remain strong and united. Our roots allow us to draw strength from our ancestors and pass down our cultural heritage to future generations.

Like the roots of papaya trees, which grow deep and spread wide, our cultural connections are resilient and far-reaching. They anchor us firmly to our heritage and enable us to thrive, no matter the challenges we face or the miles we are set apart. If you have ever tried to remove a tree, you know that despite it being cut down, the roots remain anchored firmly in the earth. This deep connection ensures that our traditions and values are nurtured and passed on to our children and grandchildren. The deeper the roots, the stronger the tree stands against storms. The storms we have endured as Indigenous Peoples are the colonization of our Lands and all the systems that colonization introduced.[1] They sought to uproot and erase our traditions but forgot that our roots anchored us strongly to our Lands. Our roots are the reason why despite all the settler colonial tactics our ancestors, grandparents, and parents endured, Indigenous cultures are still alive and thriving in modern times.

Different assimilation and eradication efforts were introduced throughout Latin America against Indigenous Peoples. For example, in Mexico, the government implemented policies aimed at eliminating Indigenous cultures, such as the suppression of Indigenous languages. In El Salvador, similar efforts were made to suppress Indigenous identities, including banning traditional clothing in addition to prohibiting Indigenous languages. In Guatemala, they tried to eradicate Indigenous cultures through violent campaigns and forced displacement. In Peru, they implemented policies that promoted European customs and marginalized Indigenous Peoples' traditions. The list of settler ideologies and colonial tactics could keep going on. However, as Indigenous Peoples, we continue to thrive despite all the assimilation techniques, forced displacement, and everything we've been through.

Today, we understand that in order to maintain this strength and resilience, we must also nourish our roots, both in our ancestral Lands and the newfound homes we have been displaced to. This means that no matter where we are, whether on our ancestral Lands or in new territories, we must safeguard our cultural heritage. Each of us is like a tree in this vast forest, and together we create a living testament to what it means to be Indigenous. Like a forest, we support each other, share resources, and thrive through our interconnectedness. Our roots run deep, carrying the wisdom and traditions of our ancestors, grand-parents, and parents. Together, we stand resilient against the winds of extreme change and adversity. The dense canopy and thick underbrush that make up our Indigenous cultures made it incredibly difficult for them to efface our identities. The lack of clear paths and the presence of plant and animal spirits further hindered their progress. While many of our ancestors did succumb to the diseases, violence, and genocide they introduced in our Lands, we are still here today. We are the forests they couldn't clear.

Forests hold deep spiritual significance. The forest is interwoven with our cultural identity, traditions, and way of life. My elders always told me that we are deeply rooted and connected to every living and non-living thing within it. The forest is not here just to nurture and sustain us; we must also protect and respect it. This symbiotic relationship is at the core of our existence and responsibilities as human beings and we as Indigenous Peoples continue to carry today. While some people have lost their connection with nature, most Indigenous Peoples have not. Modern society can learn valuable lessons from Indigenous practices, particularly in terms of sustainability and stewardship. These are two things deeply tied to our spirituality. Our Indigenous spirituality is something Western religions today cannot explain or truly comprehend. Still, one thing that we can mention about our spirituality is that we know that communities can work toward a healthier planet by adopting a more respectful and symbiotic relationship with nature. Embracing these practices can help mitigate environmental degradation and foster a deeper appreciation of the natural world. We are the forests for our future generations, because we provide them with the wisdom, resources, and the spiritual grounding they need to thrive. Just as the forest nurtures its inhabitants, we nurture, and guide, our descendants.

For some Indigenous communities, like my own, a forest is a vital source of food, medicine, and materials for shelter and tools.[2] Unfortunately, in modern times, we are experiencing climate change that threatens our identities by severely jeopardizing our Lands. Climate change is altering ecosystems, endangering wildlife, and displacing communities. Climate change impacts our Lands by causing more frequent and severe natural disasters such as wildfires, floods, and hurricanes.[3] These events can devastate agricultural productivity, leading to food shortages and economic instability. Rising sea levels threaten to submerge coastal regions, displacing millions of people and destroying habitats. Climate change is also having a severe impact on human health. Warmer temperatures and rising seas can contribute to the spread of infectious diseases, while extreme weather events can cause serious injuries and fatalities. This environmental crisis erodes cultural heritage and traditional ways of life deeply connected to the Land. The world is failing to address climate change. Many political leaders claim climate change is a hoax because they are widely disconnected from their environments and live in an upper-class bubble removed from the most serious impacts. This widespread denial hampers efforts to employ effective policies and take collective action. As a result, the situation continues to deteriorate, putting countless environments, lives, and cultures at risk.

The continued encroachment on our Lands is hindering our resistance and resilience, displacing our communities further and forcing our roots into foreign Lands. As we lose our connection to our ancestral Lands, our cultural heritage and identities are at risk of erosion. While global advocacy organizations place climate change front and center, not much is being done to remediate its impacts and long-term effects. Climate-induced migration is rarely discussed within immigration and climate change discourse. We cannot continue to ignore how climate change is forcing millions of people and animals to leave their ancestral

Lands. They are leaving because their homes are inhabitable or because they are experiencing extreme conditions that threaten their well-being. Climate change is a challenging topic to address due to the political complexities of attributing migration to climate change directly. Migration is driven by a multitude of factors, including economic hardship, political instability, and social dynamics. These factors obscure climate change's direct impact and have already created divisions within communities, fueling xenophobic narratives against displaced people. This continues to lead to tensions and hostility toward migrants. It is imperative that we address the root causes and foster understanding to combat harmful stereotypes that continue to fuel xenophobic sentiments within the migration discourse. Economic hardship, political instability, social dynamics, and climate-induced migration are intricately linked, creating a web that displaces many Indigenous Peoples among other communities. Despite this interconnectedness, global and national agendas often fail to adequately support these communities due to a lack of cohesive migration views. International and national policies must recognize and address migration issues' multifaceted nature to provide comprehensive solutions.

It is interesting to note that Global North countries like the United States and Canada, which have some of the strictest immigration policies, are often the least proactive in addressing the root causes of migration, particularly within climate-induced migration. These nations have the resources and capabilities to take the lead in mitigating climate change's impacts but often fall short of taking meaningful action. Instead, they focus on tightening their borders, which exacerbates the plight of those forced to migrate due to environmental crises. We get politicians from those countries coming to our Lands and telling us not to come instead of taking action to reduce their greenhouse gases, eliminate mining projects they own on our Lands, or return our Lands, which their monocultural corporations are destroying. This hypocrisy is glaring. They lecture us on migration while contributing significantly to environmental degradation that forces us to leave. This is why we need global platforms that advocate for climate change, such as the Rio Conventions, to press for accountability from the largest emitters of greenhouse gases. The Rio Conventions are international treaties that address global environmental issues and aim to combat threats to biodiversity, climate, and desertification.[4] These conventions include the Convention on Biological Diversity (CBD), the United Nations Framework Convention on Climate Change (UNFCCC), and the United Nations Convention to Combat Desertification (UNCCD).[5] These three conventions were established as a direct outcome of the 1992 Earth Summit that was held in Rio de Janeiro. The summit marked a pivotal moment for global climate governance, bringing together nations to address pressing environmental challenges. These platforms can help ensure that the nations most responsible for environmental degradation take meaningful steps to mitigate their impact. By holding these countries accountable, we can work toward more equitable solutions for those displaced by climate-induced factors. To create real change, countries from the Global North must acknowledge their role in perpetuating the crisis. They must collaborate with affected Indigenous communities to develop sustainable and equitable solutions. Moreover, these policies should be tailored to the specific needs of Indigenous communities, considering the different contexts and Indigenous ways of knowing. We also need governments to provide financial and technical support to ensure that Indigenous communities dealing with displacement due to climate change can effectively cope through climate mitigation and adaptation strategies.

In our Lands, and in the diaspora, the duality we face as Indigenous Peoples when it comes to climate change comes from how deeply connected we are to the Land. For example, even in the diaspora, extreme weather events like heat waves are important to highlight since many displaced Indigenous Peoples from Latin America are employed in this country's agricultural sector as farmworkers.[6] Several surveys have been conducted to account for the number of Indigenous farmworkers in the United States. For example, findings from the National Agricultural

Workers Survey (NAWS) conducted from 2019 to 2020 revealed that 6 percent of all farmworkers identified themselves as Indigenous.[7] Many Indigenous farmworkers speak neither English nor Spanish, making it difficult for them to understand emergency communications. This disconnect highlights the necessity for multilingual alert systems and culturally competent outreach to protect these vulnerable communities effectively. In this situation, language barriers make our displaced Indigenous relatives more susceptible to harmful climate change and extreme weather events.

In 2021 the Indigenous Farmworker Study (IFS) identified that 165,000 Indigenous agricultural workers,[8] primarily from Mexico lived in California. Most Indigenous farmworkers in the IFS survey spoke Mixteco, a language native to Mexico's Oaxaca region. The second largest population of Indigenous farmworkers spoke Zapoteco, another Indigenous language from the same Mexican state of Oaxaca. Their linguistic diversity highlights the cultural richness and unique challenges our Indigenous communities face in the agricultural sector. This is why language is a vital element to highlight.[9] For many years, and even in some locations today, Spanish has been viewed as the primary language of Indigenous farmworkers. Unfortunately, this results in exclusions from information, protections, services, and resources, thus amplifying their vulnerabilities and leading them to dangerous, or even deadly, situations. It is crucial to recognize the distinct cultural and linguistic needs of Indigenous farmworkers rather than treating them as a homogeneous group under the broad label of Latinx. Protective services and safety programs must be tailored to address their unique circumstances to ensure their health and well-being. Failure to do so further marginalizes these communities and puts their lives at greater risk.

Climate change exacerbates extreme weather patterns, leading to more frequent and intense heat waves that pose significant risks to farmworkers' health and safety.[10] Moreover, the increasing prevalence of wildfires further endangers their living and working conditions, making it critical to address the specific needs of these communities. Providing adequate resources and support to Indigenous farmworkers is essential to mitigate climate-related challenges and situations that often lead to deaths. For many years, wildfires have devastated the lives of farmworkers who continue to perform their jobs under dire circumstances. For instance, in 2009, photos taken from central Washington's Yakima Valley showed farmworkers picking apples while ash rained down from the sky.[11] This stark image underscored the urgent need for improved protections and resources for farmworkers exposed to hazardous conditions. It serves as a poignant reminder of the vulnerabilities faced by Indigenous farmworkers amid the dual threats of climate change and systemic neglect that need to be highlighted and advocated for more effectively. Due to these conditions, the workers were breathing in fine particulates known as PM2.5, which, when inhaled, travel deep into the lungs and seep into the bloodstream, resulting in worsened respiratory problems, impaired cognitive function, and strokes.[12] Upon conducting more research into why farmworkers did not evacuate and continued to work, it was found that language barriers prevented them from receiving evacuation alerts from authorities.

In addition to wildfires, heat waves exacerbated by climate change also create extremely harmful working conditions for Indigenous farmworkers.[13] These conditions not only threaten the workers' health but also their livelihoods, as they often lack access to other employment opportunities. In 2021 Oregon became the first state in the West to approve protective emergency heat rules for workers, setting a precedent for other states to follow.[14] Two years later, in 2023, Washington state had a permanent heat rule effective July 17. This law states that as soon as it reaches 80 degrees outside, employees must be permitted and encouraged to take preventative cool-down breaks in the shade or utilize other methods provided by their employers to reduce body temperature.[15]

But enforcing this heat rule presents several challenges, particularly in ensuring compliance among employers. Agricultural operations may lack the oversight necessary to ensure breaks are adequately provided and workers are informed of their rights. Additionally, language barriers can hinder effective communication about the revised regulations, leaving many Indigenous farmworkers unaware of the protections available to them. However, as displaced Indigenous Peoples, we need to start and continue to advocate for our relatives employed in the agricultural sector. This includes raising awareness about heat regulations and ensuring they are enforced properly. It is essential to acknowledge the many Indigenous-led community organizations that are already doing this work and playing a critical role in ensuring that our communities are still safe even when we are away from our Lands, especially since we still face dire climate change impacts and conditions in the diaspora. There are Indigenous-led communities such as the International Mayan League that advocate for Indigenous languages in many different services, including immigration.

Additionally, to address these language barriers and challenges, it is also essential to develop targeted outreach programs that effectively communicate workers' rights and safety regulations in multiple languages. Collaboration with Indigenous-led organizations can enhance vital information dissemination and provide culturally relevant support. Increased funding for regulatory agencies can improve oversight and enforcement, ensuring employers comply with heat protection rules and safeguard Indigenous farmworkers' well-being. Our responsibilities to protect our planet and heal our Lands do not end back home; it continues with us in the diaspora. Therefore, Indigenous Peoples living on their ancestral Lands and those who have been displaced to these Lands should be protected and supported by climate policies. Governments and countries should include both Indigenous Peoples native to these Lands and those who have been displaced within any policy reforms and initiatives to address climate change. It is very important that Tribal and Indigenous sovereignty are also placed at the forefront of climate policy consideration. We urgently cannot ignore Indigenous Peoples who are displaced, since they face climate change impacts both at home, in their ancestral Lands, and in the diaspora. As we have discussed, Indigenous farmworkers and the impacts they face that are harmful and deadly are a prime example of how we as displaced Indigenous Peoples continue to experience this duality in climate change impacts.

This is why it is important to raise the alarm that Indigenous Peoples need to be involved in policymaking processes by guaranteeing their representation on climate policy committees and advisory boards. If both displaced and nondisplaced Indigenous Peoples are considered in climate policies, these policies can become holistic since they start considering what happens when these impacts aren't addressed, such as forcible uprooting. Acknowledging the imbalance between the support Indigenous Peoples are receiving and the climate change impacts they are experiencing can help create fair and effective policies. These policies will also help address the needs of those most affected by climate-induced migration. Climate-induced migration is reactive, forcing Indigenous communities to leave their ancestral Lands on short notice. The abrupt nature of displacement places immense pressure on these communities to acclimate swiftly to changing environments. Immediate support and resources are crucial to help them navigate these sudden changes and establish new livelihoods. Climate-induced relocation is thus widely expected to become an adaptive response, especially for Indigenous communities who are already vulnerable and often living in environments severely impacted by climate change.[16] These communities face unique challenges due to limited access to resources and infrastructure, making targeted interventions essential. By prioritizing their needs, we can ensure relocation efforts are both sustainable and respectful of their cultural heritage. This adaptive response requires proactive planning and support for a smooth and sustainable transition. Therefore, policymakers must incorporate inclusive strategies that prioritize Indigenous populations' resilience and well-being in the face of climate-induced migration.

Moreover, these policies should be tailored to Indigenous communities' specific needs, considering equity and justice. Equity and justice must be at the center of any climate action, solutions, and policies. Without these principles, Indigenous communities are at risk of being further marginalized and disproportionately affected by the consequences of climate change. Their voices must be heard, and their rights must be respected in all decision-making processes. Governments should also provide financial and technical support so Indigenous communities can effectively cope with migration challenges. The Global North, responsible for most historical greenhouse gas emissions, must acknowledge its role in driving climate change. Climate change's adverse impacts disproportionately affect the Global South, exacerbating existing vulnerabilities and displacing communities.[17] Acknowledging this imbalance is crucial for creating fair and effective policies that address the needs of those most affected by climate-induced migration.

The slow onset of environmental changes, such as desertification or sea-level rise, makes it difficult to pinpoint a single cause for migration decisions. Consequently, policymakers struggle to create effective strategies that address these issues' intertwined nature. Governments and policymakers may be hesitant to confront these issues due to the significant economic and social implications of large-scale displacement. Climate-induced migration forces Indigenous Peoples to leave their homes due to sudden or long-term changes in their local environment and their Lands. Environmental changes make it difficult or impossible for individuals to sustain their livelihoods, leading them to seek safer and more stable living conditions elsewhere. This type of migration highlights the urgent need to address climate change impacts on populations, such as Indigenous Peoples who never want to leave their Lands. During the UNFCCC Conference of the Parties (COP) 26, we all recall the powerful message Tuvalu's prime minister, Simon Kofe, delivered standing knee-deep in the water.[18] He showed others about the lived experiences many Indigenous Peoples experience, especially coastal Indigenous Peoples and communities. Tuvalu, like many small islands, is affected by rising sea levels. This island is becoming uninhabitable because as the sea level rises, part of their island is pushed underwater. This small island nation faces the real threat of forced evacuation. This will result in them losing not just their Land, but also their cultural heritage and identity. Simon Kofe's poignant speech highlighted the urgent need for global action to address climate change and protect vulnerable communities like Tuvalu.[19] Yet, since that speech was made three years ago, not much has been done globally to address climate-induced migration. Unfortunately, the Intergovernmental Panel on Climate Change (IPCC) projects Tuvalu could be the first country to sink underwater. This stark projection underscores the imminent danger faced by small island nations. It highlights the urgent need for comprehensive international strategies to mitigate climate change and support affected communities. Immediate action is necessary to prevent the loss of entire countries and their cultural identities. Imagine being forcibly displaced from your Lands because the ocean swallowed them due to sea-level rise. The emotional and psychological toll of losing one's home and cultural heritage is unimaginable. This is the devastating reality that many small island nations, like Tuvalu, face.

In addition to rising sea levels, Tuvalu and many other small islands suffer from coastal erosion, droughts, coral bleaching, and storm surges.[20] While Western science is still trying to catch up to climate change and find solutions, Indigenous Peoples are already seeing the effects. They are forced to adapt to survive. Much like many other Indigenous communities, Tuvalu residents formulate their own assessments of changing ecological and climatological patterns.[21] They rely on traditional knowledge and observations to understand ocean tides and weather changes. Traditional knowledge and lived experiences are essential for engaging in global discourses on climate change and finding sustainable solutions. We must move away from solely focusing on Western science as it is clearly not leading us to equitable and just climate solutions. Instead, we should recognize the strength of Indigenous knowledge systems and the value they add when addressing climate change. By integrating these diverse perspectives, we can create more comprehensive and inclusive strategies to combat the global climate crisis. Indigenous science can create synergies with Western science, leading to holistic solutions that address the complexities of climate change. The lived experiences of Indigenous Peoples complement these synergies, offering practical insights and adaptive strategies that Western approaches may overlook. Together, these combined knowledge systems can pave the way for more effective and sustainable climate action. While settler colonial frameworks now dominate the world, making it difficult for Indigenous communities to participate, we are not giving up and continue to advocate for climate justice and the healing of our Lands. Indigenous voices must be amplified and included in decision-making processes to ensure that climate solutions are just and equitable. By standing united, we can push for policies that respect and integrate Indigenous knowledge, ultimately fostering a healthier planet for all. It is imperative to have the power to change mainstream discussions and hold governments accountable for their actions, especially those in the Global North.

My ancestral Lands and the Lands where I currently reside face dire threats to their forests. Climate change is causing pests, diseases, and extreme conditions in our forests. For example, in the Pacific Northwest, where I live now, Western red cedar, an important relative of Coast Salish, is at risk. Due to shifting climate patterns, the Western red cedar faces increased vulnerability to root rot and other pathogens. Western red cedar is used to build canoes, longhouses, and totem poles. Its bark is woven into baskets, mats, and clothing. However, due to climate change, Western red cedars are suffering dieback. This is a condition in which a tree or plant dies from the tip of its leaves or roots inward. This dieback threatens our forests' biodiversity and endangers the Coast Salish people's cultural practices and traditions.

In Mexico, howler monkeys face significant threats due to deforestation and habitat loss. These primates rely on the rainforest for survival. As the tree numbers dwindle, the monkeys' food sources and shelter diminish. According to reports from May 2024, howler monkeys have been dropping dead from trees in Mexico's southeastern tropical forests amid drought and heat waves that have sent temperatures soaring.[22] Extreme weather conditions are exacerbating the already critical situation for these primates, further highlighting the urgent need for sustainable environmental practices. Howler monkeys serve as symbols in our traditions and hold profound cultural significance for many Indigenous communities in Mexico. Their presence in the rainforest is deeply intertwined with our spiritual beliefs and practices. The loss of howler monkeys due to deforestation and climate change thus represents not only an ecological crisis but also a profound cultural loss for these Indigenous Peoples. Both regions underscore the global impact of climate change on diverse ecosystems and cultures. The loss of these trees would not only impact the environment, plants, and animals, but also cultural practices passed down through generations.

What does this mean for Indigenous Peoples if we are as resilient as forests? This means that our resilience as Indigenous Peoples will be tested, just as our forests are. We must draw on our traditional knowledge and Indigenous science to protect and restore these vital ecosystems. However, power dynamics have not allowed us to lead global climate actions and invaluable, Indigenous knowledge has been sidelined in climate discussions. This exclusion hinders the global community from benefiting from sustainable practices honed over centuries. While our connection to the Land guides us in finding sustainable solutions to these challenges, we need our rights to be solidified. This is so we can do more for the world. Recognizing and respecting Indigenous sovereignty are essential for effective climate solutions. To accomplish this, Indigenous Peoples must address colonial systems that are exacerbating climate change impacts. One of these colonial systems is carbon capitalism.

Carbon capitalism, which drives fossil fuel extraction and consumption, directly opposes sustainable ways of living.[23] Carbon capitalism is a term used to describe the economic system that relies on fossil fuel production and consumption for growth and profit.[24] This system often leads to environmental degradation and the disruption of traditional Lands, threatening the livelihoods and cultural heritage of Indigenous communities. It perpetuates socioeconomic disparities, making it difficult for Indigenous Peoples to maintain their way of life. The prioritization of profit over sustainability undermines efforts to implement Indigenous-led conservation and climate initiatives. Carbon capitalism is increasingly accepted as directly connected to climate change. It's also causing deaths, diseases, displacement, and destruction of the environment and people's lives. The profit-driven environmental crises have led to catastrophic impacts on vulnerable communities. Indigenous Peoples are on the frontlines of these changes and face heightened risks and challenges in preserving their Lands and cultures.

Carbon capitalism started with the materiality of coal, which fueled the Industrial Revolution and set the stage for the extensive use of fossil fuels. The extraction and burning of coal transformed economies but also began a long history of environmental degradation, which laid the groundwork for an economic system heavily dependent on nonrenewable resources, contributing to the climate crisis we face today. Coal supported the rapid growth of cities throughout the Global North. But at what cost? Coal mining has led to significant environmental destruction, including deforestation, soil erosion, and water pollution.[25] The process releases harmful pollutants into the air, contributing to respiratory problems and acid rain. Abandoned mines often leave behind toxic waste, which can seep into local water supplies, posing long-term health risks to nearby communities. The health impacts of pollution and hazardous working conditions can reduce community well-being and increase healthcare costs. Communities near coal mines often face economic instability because they become reliant on a single industry, which can lead to boom-and-bust cycles. The influx of workers and the subsequent decline when mines close can strain local infrastructure and social services. Our reliance on fossil fuels is like a house built on an unstable foundation. As the years pass, the structure becomes more and more unstable until it collapses. The carbon capitalism of the Industrial Revolution is just as unstable, and environmental degradation ultimately leads to the current climate crisis.

Another form of capitalism that jeopardizes our environment is climate capitalism. Climate capitalism refers to the intersection of climate change and economic systems, involving the assessment of the political dilemmas posed by climate change.[26] It challenges corporate power, addresses energy consumption, and seeks solutions within the capitalist framework.[27] However, it often falls short of addressing the root causes of environmental degradation and may prioritize profit over true sustainability. Many attribute carbon markets to this form of capitalism. Carbon markets allow companies to buy and sell carbon credits, theoretically encouraging emissions reductions.[28] However, critics argue that this system often enables businesses to continue polluting while paying for the right to do so rather than making substantial changes to reduce their environmental impact. This is a way for them to offset their carbon emissions by buying credit instead of reducing greenhouse emissions. Unfortunately, carbon markets are not implemented equitably, and Indigenous Peoples often face forcible displacement from their Lands.[29] This displacement disrupts their traditional ways of life, leading to the loss of culture, heritage, and livelihoods. Carbon market projects often exclude Indigenous knowledge and practices, undermining sovereignty and self-determination in some cases. As a result, while carbon markets may present a facade of environmental progress, they perpetuate social injustices against Indigenous communities. However, carbon markets are one of the climate solutions that continue to be presented globally. Proponents argue that, if properly regulated, these markets can drive innovation and investment in cleaner technologies. Carbon markets remain a prominent feature in international climate policy discussions and agreements despite their flaws.

However, given the holistic nature that Indigenous knowledge and science encompass, we must not dismiss carbon markets as an ineffective solution. We must also analyze the successful examples of carbon markets supporting Indigenous communities. One such example is the Suruí Forest Carbon Project in Brazil.[30] This project has successfully integrated Indigenous knowledge to manage and protect forests while generating carbon credits.[31] Indigenous communities across the Amazon practice some form of "Buen Vivir" or "Sumaj Kausay," roughly translated as "Living Well," for governing their natural resources, based on communal decision-making.[32] This approach emphasizes harmony with nature and collective well-being, contrasting sharply with profit-driven motives. By integrating these principles into carbon market projects, there is potential for more sustainable and equitable outcomes that respect and uplift Indigenous practices. The Suruí Forest Carbon Project in Brazil reduced deforestation dramatically during its first five years of operation (2009–2014), but a surge in deforestation followed the discovery of large gold deposits in 2018.[33] This unfortunate turn of events underscores the challenges of carbon market projects when economic incentives such as mining come into play. Nonetheless, the early success of the Suruí Forest Carbon Project demonstrated the potential for integrating Indigenous knowledge into environmental initiatives. The project generated 299,895 carbon offsets certified under the Verified Carbon Standard (VCS) before being suspended.[34] This is equal to removing sixty-four thousand fossil fuel–driven cars from the road for a year. Such achievements highlight the significant impact well-designed carbon market projects can have on the environment and Indigenous communities. Another example of a carbon market that led to positive outcomes for Indigenous communities is the Afognak Carbon Project in Alaska.[35] The Afognak Forest Carbon Project results from more than a decade of dedicated efforts, in cooperation with the Rocky Mountain Elk Foundation and the American Land Conservancy, to permanently preserve a truly special ecosystem.[36] The project is also home to Alaska's largest elk population, adding to its ecological significance. Combined with conservation efforts, the Afognak project mitigates climate change while safeguarding critical wildlife habitats. The Native Corporation collaborates with partners to manage forestlands sustainably and benefits from carbon offset revenues.[37] Through initiatives led by the Native Corporation, Land disruption and greenhouse gas emissions from timber logging have been significantly decreased. This helps protect the ecosystem and strengthens the community's role in managing their natural resources. These projects showcase how, with proper inclusion and respect for Indigenous rights, carbon markets can support both environmental and social goals.

While these projects demonstrate some successful cases, it is also imperative to listen to Indigenous communities' critiques of carbon markets. While these projects can help us protect our Lands and forests, if the companies investing in them fail to do their part in combating climate change, our Indigenous communities will still be negatively affected. Indigenous communities often argue that carbon markets lead to the commodification of their Lands and undermine their sovereignty. They also point out that these markets can result in Land grabs, displacement, and the disruption of traditional practices and lifestyles. Moreover, such projects may prioritize corporate profits over Indigenous Peoples' well-being and environmental stewardship. For example, some REDD+ projects ignore Indigenous communities' rights, tenure, and engagement.[38] Consequently, these projects encouraged and resulted in displacement, conflict, corruption, impoverishment, and cultural degradation. Additionally, Indigenous Peoples are concerned that tree plantations or monocultures will replace their Lands, which will negatively affect their ecosystem's integrity, biodiversity, and sociocultural value. This will ultimately increase carbon stocks, but these monocultures can disrupt the natural balance of the ecosystems. This will lead to the loss of flora and fauna crucial to the community's way of life. Furthermore, non-native species can have negative impacts on the local environment. These issues highlight the need for genuine and respectful collaboration with Indigenous Peoples to ensure that carbon market initiatives do not perpetuate harm. Indigenous communities often ask me if I recommend their communities consider the carbon market. My response is that they need to make those decisions for themselves. Indigenous communities should start by conducting thorough research on the potential impacts, benefits, and harms of carbon markets. Most importantly, they should engage with experts in environmental law, Indigenous rights, and sustainable development to receive valuable insights on protecting their rights and Lands. Additionally, it is crucial to involve the entire community in discussions and decision-making processes to ensure that their collective voice and traditional knowledge guide the way forward. I cannot ignore the reality that these are usually the only economic opportunities our communities receive. I also cannot ignore the harsh reality that if we do not hold corporations, companies, and governments accountable for their greenhouse emissions, Indigenous Peoples, and their roots, will continue to face climate change impacts. Therefore, carbon markets pose a double-edged sword, and Indigenous Peoples need to make the best decision in order to preserve their cultures and communities. In spite of the impact of climate change and carbon capitalism on our communities, one thing is certain: We will never lose sight of the roots of our Indigeneities. Indigenous identities are not just cultural but deeply political. They represent a history of resistance, resilience, and the ongoing struggle for rights and recognition. This dual nature makes them powerful symbols of both heritage and activism. In Mexico, Indigenous identity is a political identity because it involves a constant fight for Land rights, autonomy, and the preservation of languages and traditions against systemic marginalization. Indigenous communities often engage in activism to demand governmental recognition and protection of their cultural and territorial integrity. This struggle highlights the intersection of cultural heritage and political advocacy in their efforts to secure justice and equality. Indigenous activists in Mexico face numerous challenges, including violent repression from state forces and paramilitary groups.[39] They also encounter legal and bureaucratic obstacles that hinder their efforts to claim Land rights and protect their communities. Systemic discrimination and marginalization make it difficult for their voices to be heard and their rights to be recognized.

In El Salvador, like in Mexico, the Indigenous identity is political because it involves fighting for the recognition of cultural heritage and the preservation of ancestral Lands. Indigenous communities there advocate for their rights in the face of historical injustices and ongoing marginalization. They strive to protect their traditions, languages, and territories from encroachment and exploitation, making their identity a powerful form of resistance. In Mexico, the Zapatista movement established autonomous communities that prioritize Indigenous governance.[40] In El Salvador, the Nahua work tirelessly to reclaim their ancestral Lands and revitalize their cultural heritage.[41] They advocate for legal recognition and protection of their rights, aiming to preserve their unique way of life for future generations. Both Indigenous activists in Mexico and El Salvador face severe repression and violence from state and nonstate actors, posing significant threats to their safety and well-being. In Mexico, activists frequently encounter legal and bureaucratic barriers that complicate their quest for Land rights, while in El Salvador, the struggle often centers on overcoming historical injustices and protecting ancestral Lands from exploitation. Despite these regional differences, both groups share the common goal of preserving their cultural heritage and achieving political recognition and autonomy and are constantly fighting for climate justice.

Sometimes in the diaspora, our climate justice fight must continue because climate change continues to displace us. Throughout Latin America and the Caribbean, many Indigenous Peoples put on small backpacks and straw hats, saying final goodbyes to their families as they are obliged to leave their homes. The spread of crop diseases, such as fungi, forces them to sell their Lands. They are forced to leave their homeland and migrate to other countries to find a better life. This displacement has a lasting impact on their culture and community and highlights the need for greater action on climate change. Even the term *climate refugee* has not been clearly defined within asylum laws and regulations. This lack of legal recognition and support leaves displaced people without the necessary resources to rebuild their lives. Asylum laws and policies supporting climate refugees are rare in countries where they are forced to migrate.[42] Most are left to navigate complex immigration systems, often applying for either affirmative or defensive asylum, which do not explicitly recognize climate-related displacement.

Affirmative asylum is sought by individuals already in the country who apply for protection through the asylum office before they are placed in removal proceedings.[43] Defensive asylum, on the other hand, is requested as a defense against removal when individuals are already in deportation proceedings.[44] Both processes can be arduous and do not directly address the unique challenges climate refugees face. Furthermore, most migrants can only seek asylum in the United States if they prove they have a credible fear of persecution in their country. This means that those displaced by climate change often struggle to meet the stringent requirements, as the concept of "credible fear" is typically interpreted in the context of war, violence, or political persecution rather than environmental devastation. As a result, many climate refugees are denied asylum and left in limbo. They are unable to return to their home countries while also not being legally recognized in their new ones. Consequently, many are left in a precarious situation, struggling to find safety and stability in a new Land. This gap in protection underscores the urgent need for the international community to redefine asylum laws to include those fleeing environmental crises.

As the daughter of a climate refugee, I have experienced what climate refugees in the United States go through. While my father was displaced due to the genocide in Central America, my mother left her Lands because her parents were struggling. She comes from a family of nine children, and her parents taught them how to live off the Land. They stewarded a small orchard of papaya, mango, and guava trees, which provided both sustenance and a modest income. However, in the 1980s, when they started facing severe climate impacts on their small orchards and Lands, as records demonstrate, this was the decade when Oaxaca experienced the worst droughts. The persistent lack of rain decimated their crops, making it impossible for them to sustain their livelihood. This forced my mother and her family to decide to leave their home in search of better opportunities. My father had already been welcomed and granted refuge in my maternal pueblo. My mother bravely led us into our family's future as she hoped to find stability and a chance to rebuild our lives. Their journey was filled with challenges, but her resilience and determination united the family through hardships. Upon arriving in the United States, my mother and father faced numerous challenges. They struggled with language barriers, which made it difficult to find work and integrate into this new lifestyle. Additionally, securing affordable housing was a constant battle. They often lived in cramped, substandard conditions while working tirelessly to establish a stable life.

Like many immigrants, my parents did not have an easy journey to this country. As young asylum seekers and climate refugees, the hardest thing was to say goodbye. My mother told me the most difficult goodbye was to my sister, who was just a baby at the time. The journey to the United States was fraught with dangers and uncertainties, making bringing a young child, especially a baby, impossible. She left my older sister in my grandparents' care, hoping to reunite with her once they established a more stable life in the United States. Maternal love is beautiful and deeply cherished in my Indigenous family and community. However, this maternal love drove my mother to seek opportunities that were nonexistent in our Land due to climate change. Her unwavering commitment to providing a better future for her children pushed her to make unimaginable sacrifices. She wanted to support her eight siblings, parents, and daughter in adversity. Society passes judgment on mothers who leave their children behind, but for many displaced Indigenous women, there are no better options. The stories of the sacrifices of Indigenous mothers should not remain unheard or untold. Mothers making such sacrifices suffer an immense emotional toll. They endure the heart-wrenching pain of separation, the constant worry about their children's well-being, and the guilt that haunts them daily. Even in moments of happiness, they can only think about seeing their children again. These sacrifices leave an indelible mark on their hearts, a testament to their boundless love and resilience.

If their children are old enough to travel safely, some mothers bring them along. Still, each step of the journey is filled with peril. The constant fear of losing a child to trek dangers is a burden no parent should bear. Sometimes on these journeys, a mother will sacrifice her life to save her child from drowning in the Rio Grande or from dehydration. Other times, there is nothing a mother can do for their children, and these losses destroy them deeply. During their journey, my mother witnessed other mothers lose their children. These heartbreaking scenes served as a constant reminder of the risks involved in a better life. Indigenous Peoples do not come to the United States to steal jobs or exploit this country's economy. More importantly, the US economy has overexploited and thrived at the expense of our Indigenous Lands. These are the insights we need to illuminate and spread instead of false narratives that feed xenophobia in this country. My mother was separated from her Lands and her family, including her daughter. Family separation has profound and far-reaching effects on many displaced Indigenous Peoples who are uprooted abruptly. When children are abandoned, they can feel lost; their emotional development can suffer, making it difficult for them to form healthy relationships in the future. No mother wants that for their children, but for many Indigenous mothers, this is the only option they have for survival.

My parents' journey to the United States was filled with fear, longing, and laughter. When my father recounts their immigration story, he never misses telling the part where my mother accidentally sat on a short barrel cactus, a plant my mother had never seen, in Arizona. This was after jumping off the tren carguero. They had been riding the tren carguero for days, dealing with exhaustion and uncertainty. Despite the hardships, moments like these brought levity to their arduous journey. They had to hide within the tren carguero's openings and my mother wanted to sit down, hidden from visibility. My father always laughs loudly when recounting this part of their journey—I always wonder how he can handle such a high volume for an individual only five feet tall. Humor amid adversity can serve as a powerful coping mechanism. My parents' ability to laugh during challenging times speaks volumes about the strength of their bond.

It also highlights their resilience. Jumping from the tren carguero was the last thing they had to do. My parents strongly believe that one of their ancestors guided them on this dangerous journey. They had an odd encounter that they cannot comprehend to this day. During the final stretch of their journey, there was an elderly man dressed in all white and wearing a long straw hat accompanying them. He was even told to take off his hat because it was highly visible and would make it easier for the migra to spot them. The elder told him not to worry and follow his lead. My father remembers telling my mom that he felt sorry for this elder as there was no way he could jump the fence or run fast when they had to. However, to their surprise, this elder was faster than them and was so agile he didn't even meet the fence when he jumped. When it was time for their final jump over a fence that separated them from American soil, the elder told them he had done his job and gave them a reassuring nod before disappearing into the night. He couldn't have gotten far in a matter of seconds, but he vanished immediately, astounding my parents. They looked around in disbelief and they wondered if they had just been guided by an otherworldly presence. Even now, they reflect on that mysterious figure with a sense of protection.

This experience impacted my parents profoundly, solidifying their belief in guardian spirits, our ancestors. These spirits can appear as themselves in spirit form or they send someone to protect us. This encounter became a cornerstone of their narrative, a testament to the extraordinary moments that can occur in the most desperate times. This story has been passed down through our family over the years. It inspires us to believe in the unseen and trust that we are never truly alone. My mother shares that she felt at ease while the elder was around, but she gets chills from not understanding who he was or how he ended up guiding them with such ease. The mystery surrounding the elder's identity and his timely appearance continues to baffle her. It amazes my mother that out of all the people taking this journey back then, the elderly man chose to be with them. He guided them through paths that others did not take. My father even offered him some cash, but the elder refused and said he was happy to do it. This experience only deepened my mother's conviction that they were destined for this journey. They were protected by a benevolent force, a spirit of our ancestors that transcends beyond our understanding of the world.

Once in the United States, my mother sought employment in multiple jobs, tirelessly sending money back home to ensure her family would survive. From selling shoes to cleaning houses, my mother did it all. Her efforts were a testament to her strength and profound love for those she left behind. The emotional toll of being away from her family was immense, often leaving her with a deep sense of longing and loneliness. Despite her unwavering determination, the separation weighed heavily on her heart, as every day without them felt like an eternity. Her sacrifices were not just physical but also deeply emotional, marked by countless nights of silent tears and unspoken worries. Through science, we can now explain the environmental impacts my maternal family experienced that decreased and vanished their food security. According to historical weather records, Oaxaca experienced severe droughts in the following years: 1934–1935, 1940–1941, 1945–1950, 1954, 1957–1958, 1977, 1983–1984, and 1987.[45] Several of these droughts coincided with El Niño variations in Pacific sea-surface temperatures. These harsh climatic conditions exacerbated the challenges faced by my family and countless others, forcing them to seek new opportunities far from their homeland. While my maternal family and Indigenous community were not capturing this quantitative data, they experienced the impacts of these droughts. Communal milpas, essential for their sustenance, were also destroyed, plunging my family into severe food insecurity during those times. This further underscored my mother's urgency to seek better opportunities abroad.

People do not realize that climate change will continue and exacerbate environmental conditions that will exacerbate the issues of climate-induced migration in turn. Consequently, when we discuss migration, we cannot focus solely on traditional economic or political reasons. While there is growing recognition that climate change is a driving force behind another wave of migration that cannot be ignored, we must bring more attention to this form of migration. Climate-induced displacement creates new challenges for countries, but unfortunately, many governments choose to criminalize rather than address the root causes of this phenomenon. Unfortunately, as Indigenous Peoples, our communities continue to be at the forefront of climate change. We face unique challenges due to environmental degradation and Land loss. Yet instead of recognizing their rights and supporting us, many countries criminalize our migration as Indigenous Peoples. However, Indigenous migration is criminalized because of the long history of colonialism and discrimination. Indigenous Peoples have been displaced from their ancestral Lands for centuries, and their attempts to maintain their cultures and protect their families are often labeled criminal activities. This approach not only fails to address migration's root causes but also perpetuates the cycle of dispossession and marginalization.

As I am getting older and further understanding climate-induced migration and displacement, I must acknowledge that as Oaxacans, those of us who have been displaced and those who have remained in Mexico are closely knit together into dense networks that span the Mexican and US border. This is what constitutes what we call our "transnational communities." These networks provide support and resources, allowing us to maintain cultural ties and collective identity despite physical distances. They are vital for preserving our heritage and fostering resilience amid challenges. As transmigrants who are also Indigenous Peoples, our relationship with the Mexican state differs significantly from that of non-Indigenous or mestizo migrants. Our unique cultural heritage and ties to our ancestral Lands shape this relationship, adding layers of complexity to our experiences of migration and displacement. This distinct identity influences how we interact with both the Mexican state and our host countries, often requiring us to navigate additional challenges and opportunities. Our Latin American Indigenous migration is a multiethnic phenomenon that is much older than mestizo migration and has a transnational character because of strong community ties. This deep-rooted history underscores the enduring connections that bind our people across borders, reinforcing our sense of identity and unity. These long-standing networks not only facilitate the sharing of resources and support but also help us collectively navigate the complexities of transnational life. These transnational identities allow us to remain closely attached to our roots even in the diaspora, preserving our language, traditions, and cultural practices. They empower us to create vibrant communities abroad that reflect the richness of our heritage. This interconnectedness ensures that our collective identity and sense of belonging endure no matter where we are. Like papaya trees with roots that extend deep into the soil, our transnational communities grow and flourish across borders, drawing strength from the rich cultural heritage that sustains us. Just as papaya trees bear fruit in diverse environments, we, too, thrive in various Lands while maintaining our unique identity.

However, our transnational identities also call for unity and solidarity with our Indigenous relatives back home and those in the diaspora. By staying connected and supporting one another, we can collectively address challenges and advocate for our rights and well-being. This solidarity reinforces our shared cultural heritage and strengthens our communities, ensuring our traditions and identities thrive no matter where we are. It also calls for solidarity with Indigenous Peoples of the Lands to which we have been displaced. By building alliances and fostering mutual support, we can amplify our voices and work together toward common goals. This unity not only enriches our communities but also strengthens our collective efforts to preserve and promote Indigenous rights and heritage globally. Of course, there is always conflict in our communities, pueblos, and even within relationship-building efforts. This is the beauty of being distinct individuals. However, we never miss our pueblo's drama even in the diaspora, as these tensions and resolutions are part of our vibrant cultural tapestry. Aside from the entertaining chisme we receive from this drama, we also hear about new community members taking the journey because climate change continues to destroy and impact our ancestral Lands. We learn when our pueblo loses their milpas, jeopardizing our entire community's communal food source. Thus, we move quickly to support them back home, demonstrating the resilience and adaptability of our transnational identities. This interconnectedness highlights the strength of our bonds and our unwavering commitment to stand by one another, regardless of borders. This fuels our advocacy and activism to ensure that countries take more decisive actions against climate change and provide better support for climate refugees. Our collective efforts aim to influence policies that protect our Lands and secure a future for our people. By raising awareness and pushing for systemic change, we strive to safeguard our heritage and create sustainable communities for future generations.

3

Preserving Our Land:
Land Rights

Our Land protects us, nurtures us, and nourishes us. Our Land is what connects us to our ancestors and the natural world. We Zapotec people know that our ancestors lived in mountain enclaves deeply woven with our fertile valleys, creating a basket of plentiful foods. These plentiful woven baskets included our traditional foods such as corn, beans, cacao, squash, pumpkins, and chiles. While our traditional foods serve as a direct link to our ancestors, they also embody wisdom and agricultural knowledge passed down through generations. We have been taught that our traditional foods are not just sustenance but a vital part of our identity. Therefore, our foods play a major role in our cultural rituals and community gatherings. Our ancestors mastered agricultural practices in symbiosis with our environments and Lands. Their agricultural practices were not extractive or harmful.

By respecting natural cycles and biodiversity, they ensured our resources' sustainability for future generations. Their ingenuity transcends the limitations of Western scientific knowledge. The harmony and balance they achieved with nature remain unparalleled. This is because

Western knowledge is often taught in binary terms and often emphasizes compartmentalized and reductionist approaches, focusing on isolated elements rather than holistic systems.[1] This fragmentation also overlooks the interconnectedness of ecosystems and cultural practices that Indigenous knowledge respects. As a result, Western methods may fail to capture the nuanced relationships between humans, plants, animals, and the Land. This may lead to solutions that are less sustainable and harmonious. In contrast, Indigenous knowledge is deeply experiential, passed down through living practices and storytelling.[2] Our way of life embodies this knowledge and keeps it alive. This is why our past and present Indigenous practices offer invaluable lessons for sustainable living and environmental stewardship today.

While our Lands still provide us with traditional foods that our ancestors once consumed, our food baskets are not as plentiful as they used to be. I recall my grandmother's childhood stories filled with warm embraces and the scent of epazote plants. She would help her mom pick them, gently caressing the plants as a rough touch or harmful pull would make the leaves bitter. Every time I taste bitter epazote, I know that the plant was not harvested with the care it deserves, with respect for its true essence. This lesson from my grandmother taught me the importance of respect and mindfulness in our interactions with nature. Her wisdom still guides me in nurturing a deeper connection with the Land and its gifts. Plants need to be treated with respect, not just as objects. We must see them as more than food; they are a part of us as they are part of our Lands. My grandmother also told me about how she picked additional epazote with her mom as this plant cures stomach issues or helps with congested lungs. I was blessed to learn about this practice and the knowledge of many other plant relatives' medicinal and healing properties. Some of this knowledge is only passed down through family lines to keep such knowledge from falling into the wrong hands; these plants will be reorganized and commercialized massively. Time after time, we have witnessed how Indigenous medicinal plants are commercialized but the economic benefits never go to the Indigenous communities they originate from. It is unfortunate but too common in our reality. The knowledge that epazote helps with stomach issues or congestion is now well known, but our family's recipe for the tea is our heirloom. This recipe is a special blend of traditions and practices that will only be shared within our family. It is a cherished piece of our heritage that symbolizes the deep connection our family holds with our plant relatives and our Land.

Many non-Indigenous Peoples feel entitled to Indigenous knowledge. For many years, they have been taught that anything related to Indigenous Peoples can be bought and sold. Just as people believe they are entrusted to our Lands; they feel entitled to learning our cultures and ways of knowing. While some label our privacy as "gatekeeping," for us Indigenous Peoples, it is a form of protection and responsibility that we must abide by. This lack of respect ignores Indigenous knowledge's spiritual and cultural significance. This ignorance fails to recognize the unique relationship Indigenous Peoples have with our Lands and the knowledge we have gained over generations. Yes, we can work to integrate more Indigenous knowledge and science into Western knowledge systems and science, but we can never do this without Indigenous Peoples. True integration requires collaboration and respect, ensuring Indigenous voices are at the forefront of the conversation. This approach honors the depth and richness of Indigenous knowledge while promoting genuine understanding and respect for our cultural heritage. No book can serve as a comprehensive manual for integrating Indigenous knowledge or science, as this integration requires building genuine relationships with Indigenous Peoples and communities. Genuine relationships are founded on mutual respect, trust, and understanding. They cannot be created by following written guidelines alone. True collaboration emerges from ongoing dialogues and shared experiences and through fostering a deeper appreciation for the wisdom and practices nurtured over generations.

Respect for Indigenous knowledge and science requires consent from Indigenous communities. This means recognizing their sovereignty and self-determination. It also requires allowing Indigenous Peoples to determine how their knowledge is shared and integrated. Proper acknowledgment and compensation for their contributions are crucial to maintaining cultural integrity and respect. We are often forced to bear the brunt of this emotional labor that entails advocating for our rights, from our Land rights to our human rights. However, we are rarely acknowledged as collaborators and stakeholders within the realms that hold power. Our voices are often marginalized, and our contributions are minimized, leading to persistent inequity in knowledge valuation and utilization. True collaboration requires more than inclusion. It requires active respect and recognition of our expertise and leadership, especially our knowledge about our Lands. Our Lands are our identities, culture, life support, and the essence of our entire existence. Our Lands are not simply physical spaces but sacred entities that hold our history, traditions, and future. Without acknowledgment and respect for our deep connection to the Land, any attempt at collaboration will remain superficial and disconnected from the true spirit of Indigenous knowledge.

Our Land is where our first cries were heard as we were born and brought into this world. Our Land is also where our last heartbeats are felt as we transcend this world. Unfortunately, our Land is becoming ill because of human actions. These damaging human actions worsened when capitalism was prioritized over Indigenous Land rights by settler governments. These harmful actions include deforestation, extensive mining, and pollution from industrial waste. The exploitation of natural resources for profit has led to the destruction of ecosystems and the displacement of Indigenous communities. Climate change driven by excessive carbon emissions continues to threaten the health of our environment. As humans, we are not disconnected from our environments as we are a part of nature, and nature is a part of us. Our Land is our living breath, our umbilical cord that connects us to who we are, and who we will always be, as Indigenous Peoples. Just like a mother, our Land nurtures and provides us with everything we need to survive. It sustains our bodies and spirits, offering us food, water, and a sense of belonging. We must care for it with the same love and respect that we show our mothers. When I visit my maternal Lands, this is one of the many teachings that come to mind. My maternal Lands are a sacred place where my ancestors' history is written in every stone and tree. From the grecas that adorn our ancient temples to our traditional weaving, a part of our ancestors lies in everything that is a part of our existence. Walking these paths, I feel a profound connection to my ancient ancestors and my relatives, like my beloved grandmother, who have passed on. It is from here, in my maternal Lands, that I draw strength, wisdom, and guidance, especially as someone who is part of the diaspora. Being away from my ancestral Lands often leaves me feeling disconnected, but returning here renews my sense of identity and purpose as an Indigenous woman. No matter where we are in the world, our roots and heritage continue to guide and sustain us.

When I was young, my grandmother taught me how to make her healing epazote tea. We walked through the fields, sharing stories. She told me stories of her childhood and my mother's childhood. We carefully selected the freshest leaves; then she showed me how to steep them just right and mix the epazote leaves with other ingredients. The aroma filled the kitchen, and that same aroma still comforts me when I am ill. My grandmother's essence comes to me when I make this medicine. This is the power of Indigenous knowledge. It is a living, breathing connection that spans generations, healing our bodies and our spirits. This profound understanding and reverence for our ancestral wisdom sustains us and must be honored in any true partnership or collaboration.

However, as my grandmother grew older, she told me how our once plentiful baskets had disappeared. The fields, once covered with epazote, became barren. So, making our traditional medicines has become more difficult. This is due to aggressive farming practices such as the increasing use of pesticides and industrial farming methods that disrupt natural ecosystems. The commodification of our Lands has led to its infertility. The once rich soil, full of life, lost its nutrients and could no longer sustain diverse plant life like epazote. Now, we, as Indigenous Peoples, are left to defend ourselves as these monoculture plantations continue to profit off our Lands. We must fight for our Lands and fight to protect its natural resources. This is why we are the number one advocates for Land rights. I fight for Land rights for my maternal Lands but also my paternal Lands. My familial Lands will forever hold a piece of me, and I will always hold a piece of them.

My maternal Lands are flat and protected by the surrounding mountains. The forested mountains provide essential resources such as medicinal plants, clean water, and wood for building and warmth. They also serve as a natural barrier, protecting our communities from external threats such as heavy winds, flooding, and other environmental conditions that are increasing daily because of climate change. Our mountains are sacred spaces where we perform spiritual rituals and connect with our ancestors. Our Land allows us to worship gods and goddesses. We give thanks when we wake up each day. Our Land, our sacred ground, has always been the foundation of our community, shaping our culture and traditions as Zapotec people. Our Land is a source of strength, a place of refuge, and a source of abundance. Our Land serves as a labyrinth of stories and memories, some painful and some joyful. Our histories as Indigenous Peoples have been complex. During the fall season, the aroma of cempasúchil embraces the cold air. The scent reminds us that we are close to our ancestors. The smell of coffee in the morning waltzes with the scent of agave and reminds us that we are home, in our Lands.

Paternal Lands become a mirror to our past, present, and future. Our ancestral pyramids were made from stones using advanced engineering techniques. These pyramids encourage a deep connection between the past and the present. They serve as a reminder of the power in the collective knowledge of the Maya people. While the winds whisper at night, we remember the pain many of us experienced from 1960 to 1992, when our grandparents and parents were subjected to genocide against our people.[3] Yet, the smell of loroco—rich, herbaceous, and aromatic—fills the air, a comforting reminder of resilience and hope. Our native flower not only enhances our traditional dishes but symbolizes our culture's enduring spirit. The scent of loroco brings a sense of unity and continuity to our community in our Lands despite its painful past. For Indigenous Peoples from El Salvador, the sound of the water is vital. It plays the harmonies of our life's song. From the water cascading onto rocks to the water rushing from our flowing rivers, the sound cannot be forgotten. The singers of nature's ballad are the various birds inspired by the rising sun. My paternal Land reminds me of the cumbias that celebrate who we are despite our adversities. I cannot emphasize the importance of my maternal and paternal Lands in shaping my identity as a Maya Ch'orti' and Zapotec woman enough. I see these Lands as more than just places or spaces; they are the essence of my culture and history. I rely on them for strength and inspiration in my journey through the complexities of the modern world while honoring my ancestral roots.

Our Land holds profound significance in our lives. Our Lands are the epitome of our existence. A tangible embodiment of our cultural roots and identity representing our connection to our ancestors and the spiritual world where we can leave our footprints as we embark on a transformative journey into the past. However, as Indigenous Peoples, we face a distressing reality. Our Land, which has long been a source of sustenance and belonging, is increasingly being sold off to foreigners. Economies that prioritize profit over heritage and preservation consistently undervalue our Lands. The romanticization of Oaxaca, a picturesque and diverse region in Mexico, has led to a surge of foreign interest in buying vacation homes and has increased tourism in the area. Oaxaca's rich cultural heritage, stunning landscapes, and vibrant traditions attract outsiders. While this influx of tourists and investments can bring economic benefits, it also presents challenges. As an example, it can provide additional income to Indigenous families who run small businesses or sell their artesanías. On the contrary, it can also push large corporations or governments to encroach on Indigenous Lands to sell for continued building of spaces such as hotels and private properties. It disrupts traditional practices, undermines cultural continuity, and fuels dispossession and marginalization. Tourism growth in Oaxaca highlights the complex dynamics of economic development and environmental preservation.[4] It is crucial to balance promoting tourism while preserving Indigenous communities' authenticity and integrity. As Indigenous Peoples, we remain committed to safeguarding our Land and its sacred sites. We understand that our Land is not a commodity to be bought and sold, but a living entity that connects us to our ancestors and our shared heritage. We will continue to advocate for our rights and protect our identity deeply rooted in our Land. We cannot ignore the current commodification of our Lands. It is not something that faded away during the colonial period; it remains relevant today.

El Salvador, my paternal Land, has a rich and storied history and present. Unfortunately, in recent years, the country has faced significant challenges, particularly in the form of Land conflicts and violent displacement. These Land conflicts have a profound impact on the lives of many individuals today, including myself. One of the most disheartening aspects of this situation is the sale of my paternal Lands to foreigners.[5] Like in Oaxaca, in El Salvador, the sale of our territories to foreigners has intensified. This development has caused heartache but has also raised serious concerns regarding the long-term sustainability of our community and way of life. In recent years, tourism has become a significant industry in El Salvador, attracting visitors from around the world.[6] The influx of foreigners attracted to the country's natural beauty and unique cultural heritage has increased demand for Land. As a result, our territories are being sold off to individuals who may not have the same connection to our Land that we do. The sale of Lands to foreigners affects the families who were once stewards of these territories, but it also affects the broader community. Loss of communal Lands and resources harms local culture, traditions, and biodiversity, and the displacement of communities contributes to social unrest and instability.

Given how significant our Land is to us, it is no surprise that, as Indigenous Peoples, we have advocated and continue to advocate for our Land rights. To us, Land is not a physical object that can be owned; rather, it is seen as a vital part of our identity and heritage. It is the first and last chapter of our life story and serves as the foundation of our communities. Our Lands become a part of us that needs to be protected and cherished. Through profound connection, we strive to preserve and honor our ancestral territories. Despite comprising only 5 percent of the global population, Indigenous Peoples safeguard 80 percent of global biodiversity. This significant role underscores the importance of recognizing and supporting our efforts to heal and protect our planet for all of humanity. Furthermore, it is evidence enough to support Indigenous Land rights.

As a result of fighting for our Land rights, we have a distinct understanding of our traditional, communal beliefs. While some Indigenous communities still practice communal and collective rights, it is harder to embrace these within the ruling Western governmental frameworks that encourage hyperindividualism. This tension leads to a clash of values and a struggle to maintain traditional practices. The imposition of standard, individual Land ownership can undermine the communal bonds and shared responsibilities central to many Indigenous cultures. Thus, we must learn new sets of rules and regulations so that we can continue to steward our Lands despite climate change, Land dispossession, and natural disasters. Ultimately, national governments should develop legal frameworks that recognize and protect communal and individual Land rights. As an example, the Treaty of Waitangi in Aotearoa preserves the collective and individual rights of the Maori over their Lands, forests, and fisheries.[7] Moreover, a permanent tribunal was established in 1975 to adjudicate any breaches of the treaty.[8] Providing a legal framework in which communal and individual Land rights are acknowledged, the Treaty of Waitangi is a model that safeguards Maori traditional practices and values. By establishing a permanent tribunal, Indigenous communities and the government can keep track of disputes and resolve them more effectively.[9] This approach can inspire other nations to develop similar systems that honor Indigenous Land stewardship and cultural heritage.

According to Panama's constitution, five regions (comarcas) are identified based on Indigenous Peoples' rights.[10] Indigenous Peoples who live outside these regions are entitled to collective title under Law 72, which has existed since 2008.[11] The recognition of Indigenous cultures and the promotion of sustainable Land stewardship practices are crucial for preservation of Indigenous cultures. These are some examples of how Land rights have created synergy with Indigenous ways of life. As we know, Indigenous Peoples' collective rights to Lands, territories, and resources not only contribute to their cultural values, well-being, and livelihoods but also address global challenges such as climate change and biodiversity loss. As a result, Indigenous Land rights are imperative for cultural preservation and environmental sustainability and vital to climate mitigation and adaptation strategies that we as a global society must undertake to sustain life on earth.

Unfortunately, our fight for Land rights will not end until Indigenous Peoples are given the power to protect their Lands and not face any displacement from them. In many countries, Indigenous Peoples' collective rights are not recognized. This lack of recognition and procedural failure leave Indigenous communities vulnerable to exploitation and displacement. Governments must prioritize and expedite these processes to ensure Indigenous Peoples' protection and empowerment. There are times when the necessary procedures, such as Land use plans and resource mappings, are not completed, which hinders Indigenous communities from obtaining their Land rights. This often results in conflicts over Land use, unsustainable resource extraction, and environmental degradation. Without clear Land use plans, developers and corporations can encroach on Indigenous Lands without proper consent. In addition, bureaucratic delays often prolong our communities' exploitation and marginalization. Accelerating these processes is essential for securing their rightful autonomy and preserving our rights to our ancestral Lands. In cases where

Indigenous Peoples have obtained legal protection or title deeds to their Lands and resources, a lack of enforcement of laws as well as contradictory laws frequently result in a systematic denial of their rights. This legal inconsistency undermines the efforts of Indigenous communities to safeguard their territories and maintain their way of life. Therefore, it is imperative for governments to not only grant legal recognition but also ensure robust enforcement of these laws to protect Indigenous Land rights. We continue to witness the exploitation of our Lands, especially when state or foreign business entities undertake projects such as mining, logging, monocropping, or plantations without obtaining Indigenous Peoples' free, prior, and informed consent (FPIC).[12] These activities often lead to environmental degradation and displacement, further exacerbating Indigenous communities' marginalization.

Indigenous Lands encompass vast landscapes, ranging from forests, grasslands, and deserts to coastal areas and mountains. Variations in our environments demonstrate the importance of Indigenous stewardship and solidify a strong claim for Land rights. As Indigenous Peoples, we play a vital role in preserving biodiversity, mitigating climate change, and supporting other communities' cultural and economic well-being. Despite the threats our Lands face, such as Land grabbing, deforestation, and pollution, we still steward a wealth of natural resources, including forests, minerals, and water. Together, these environments are essential for sustainable development and for our lives on earth. Our Lands are home to a wide range of plant and animal species. Many species are found nowhere else in the world. By protecting these Lands, we ensure the survival of these species and contribute to the global effort to conserve biodiversity. Our forests, for example, act as carbon sinks, absorbing and storing carbon dioxide in the atmosphere. By preserving our forests, we contribute to the global effort to reduce greenhouse gas emissions and mitigate climate change. As we continue to rely on coal and other fossil fuels for energy, preserving our Land is more critical than ever. Our forests are often deemed the lungs of Mother Earth, providing oxygen and absorbing the carbon dioxide we emit. Unfortunately, when Land rights are not strengthened for our Indigenous communities, our forests, our lungs, suffer.

Indigenous communities may not have Western scientific knowledge about the greenhouse effect, but they know, based on lived experiences, that our forests breathe in the bad gases and emit the healthy gases that grant us life on this earth. When I visit my Indigenous communities, members understand why it is imperative to preserve our forests. However, deforestation has also occurred within our communities too. Sometimes, Indigenous Peoples participate in activities that may be harmful to their community because they have no other option. External pressures, such as economic hardship and lack of access to sustainable resources, force many Indigenous communities to engage in deforestation. These pressures are often exacerbated by the encroachment of large corporations and government policies that do not prioritize Indigenous Land rights. As a result, community members face difficult choices that compromise their traditional values and environmental stewardship. This is why we must not judge Indigenous communities or force them to adopt any practices or actions that they do not choose of their own free will. International frameworks like the United Nations Declaration on the Rights of Indigenous Peoples (UNDRIP) are crucial.[13] They provide a global standard for Indigenous rights protection and promote sustainable development that respects traditional knowledge and practices. Supporting these frameworks can empower Indigenous communities to make decisions that align with their values. This will ensure the preservation of their forests and way of life.

UNDRIP is an international human rights instrument adopted by the United Nations General Assembly on September 13, 2007. After a long process of negotiations and consultations among member states, Indigenous Peoples, and civil society organizations, UNDRIP recognizes Indigenous Peoples' inherent rights, dignity, and self-determination.[14] It lays out principles, standards, and guidelines to promote and protect Indigenous

Peoples' rights in areas such as culture, language, education, health, and the environment. Unfortunately, countries like the United States, Canada, New Zealand, and Australia did not adopt it initially. These nations expressed concerns over certain provisions they believed could conflict with their existing laws and policies. However, they have since endorsed UNDRIP, recognizing its importance in advancing Indigenous rights. It is important to note that UNDRIP came about because of years of advocacy led by Indigenous Peoples and communities within the United Nations. Recognizing these efforts is crucial, as it highlights the significant influence grassroots movements can have on international policy. A significant impact of UNDRIP is supporting Indigenous communities' Land rights. It has provided a framework for Indigenous Peoples to assert their traditional Land claims and seek legal recognition from their respective governments. This has led to increased protection of ancestral Lands from exploitation and development without Indigenous consent.

One of the key principles of UNDRIP is FPIC.[15] This principle requires that any decision or action that affects Indigenous Peoples' rights or interests must involve their free, prior, and informed consent; this means that Indigenous Peoples have the right to make decisions about their own development and participate in decision-making processes that affect their Lands and territories. The concept of FPIC is closely linked to Land rights. Land rights are fundamental to Indigenous Peoples' identity and way of life. They have historical, cultural, and spiritual connections with the Land, and their rights to their Land are acknowledged and protected under international law. UNDRIP recognizes Indigenous Peoples' rights to their traditional Lands, territories, and resources. UNDRIP also recognizes that Indigenous Peoples have the right to participate in decision-making regarding their Land, including the right to participate in Land-related negotiations and agreements. The application of FPIC to Land rights is crucial. This is because it helps to ensure that Indigenous Peoples' rights and interests are considered in decision-making processes that affect their

Lands. It ensures their voice is heard and their rights are protected. UNDRIP provides a framework for Indigenous Peoples' Land rights, protection, reconciliation, cooperation, and sustainable development. It encourages governments and stakeholders to engage with Indigenous Peoples respectfully and cooperatively, recognizing their inherent rights and interests. Unfortunately, such mechanisms like UNDRIP and FPIC are difficult to implement or enforce universally, especially when dealing with countries and corporations. Many governments and businesses may not fully comply with these principles, leading to ongoing conflicts and disputes over Land rights. Additionally, the lack of enforcement mechanisms and accountability measures often undermines these protections' effectiveness.

Like the papaya tree, we must not lose hope, even when it comes to global advocacy. Continuous efforts and international pressure can gradually lead to greater compliance and respect for these principles. By fostering collaboration and raising awareness, we can work toward a future where these protections are upheld. The United Nations might have limited enforcement jurisdictions, but, as Indigenous Peoples, hope drives us to continue being in these spaces. Like papaya trees that bear fruit with patience and care, our persistent efforts can eventually yield positive results. It may take time for our advocacy to bear fruit, but with determination and unity, we can see the change we strive for. Just as the papaya tree thrives under the right conditions, so too can our initiatives flourish with continued dedication and support. When we are young, we are like unripe papayas, eager and impatient for swift change. However, the reality is that meaningful transformation takes time and persistent effort. Just as a papaya requires time to mature and ripen, our advocacy efforts bear fruit through gradual, sustained commitment and resilience. However, this desire to change quickly is not a negative thing. The youth's passion for change can motivate older adults and our elders. Their wisdom and experience are complemented by the younger generations' youthful energy and impatience for progress, creating a powerful synergy. This blend of urgency and seasoned insight is essential for driving our advocacy forward and achieving lasting impacts.

When I was younger, I was impatient for change too. Over time, I learned that true progress requires patience, strategy, and collaboration. I now see the value in a balanced approach, combining youthful passion with experience to drive meaningful, sustainable change. As we grow older, it is our responsibility to mentor our youth and protect them from the harm and violence that often comes from demanding change. By guiding them and sharing our experiences, we can help them navigate the challenges more effectively. This mentorship not only empowers the next generation but ensures that our collective efforts continue to grow and thrive. I have noticed that violence and harm are enacted against Indigenous Peoples, regardless of age. This is especially true when it comes to our advocacy for Land rights. Indigenous Land and environmental defenders face significant risks and often become targets of violence and repression. Their advocacy for protecting sacred Lands and natural resources is frequently met with resistance from powerful entities. Despite these dangers, their unwavering commitment serves as an inspiration, reminding us that safeguarding our planet and our heritage is a fight worth continuing.

Unfortunately, in 2023 two hundred Land and environmental defenders were murdered.[16] Eighty-five percent of these killings occurred in Latin America.[17] This alarming statistic underscores the grave dangers faced by those who tirelessly work to safeguard our environment and ancestral Lands. Their sacrifices highlight the urgent need for Indigenous leaders to be protected and supported. As a result of intense conflicts over resources and Land in Latin America, defenders in this region have become particularly perilous.[18] It is imperative that we intensify our efforts to protect these courageous individuals who risk their lives for the greater good. Latin America is one of the most dangerous regions for environmental defenders. This is due to weak legal protections, high levels of corruption, and intense conflicts over Land and resource exploitation. The combination of these factors creates a climate in which violence and intimidation are widely used to suppress advocacy and silence those who disagree with powerful interests. Assuring the safety and effectiveness of these crucial defenders requires strengthening legal frameworks and increasing international pressure.

Mexico, my maternal Land, is one of the deadliest places in the world for Land and environmental defenders. The country has witnessed an alarming number of killings of advocates against illegal logging, mining, and other forms of exploitation. Land and environmental defenders like Tomás Rojo from the Yaqui community have disappeared only for their bodies to be discovered later. His tragic story is a stark reminder of Indigenous advocates' extreme risks. It underscores the urgent need for global solidarity and action to protect these defenders from such heinous acts. Unfortunately, our women and environmental defenders make up the majority of murders. Women who are environmental defenders face gender-based violence and discrimination alongside the dangers their male counterparts face. They are frequently targeted not only for their activism but also because of their gender, making them more vulnerable to sexual violence, harassment, and intimidation. These compounded threats highlight the need for tailored protection measures that address the unique challenges faced by women in their fight to safeguard our environment and heritage. Women who are Land and environmental defenders are perceived as a threat because they challenge deeply entrenched power structures rooted in class privileges and gender discrimination. By advocating for Land rights, they disrupt the status quo, which often benefits powerful elites. This resistance makes them prime targets for violence and intimidation, underscoring the urgency of implementing robust measures to protect these brave women. Women Land defenders like Miriam Miranda, a Garifuna community leader, risk their lives to defend their people's Land.[19] Her activism has made her a target for harassment, threats, and violence, yet she continues to stand strong in the face of these dangers. Miriam's courage exemplifies women's resilience and determination for Land rights.

Between 2012 and 2023, 2,106 Land and environmental defenders have been murdered globally.[20] These defenders are not mere statistics; they are our relatives, community healers, leaders, and teachers. They are our parents, uncles, aunts, and caretakers. When we advocate for our Land rights, we do it not for economic gains or money. Instead, we do it because our Indigenous communities are at risk as our Lands continue to be harmed. Land is essential for our cultural practices, livelihoods, and overall well-being. It provides us with sustenance, medicines, and spiritual grounding. The loss of our Lands not only threatens our cultural identity but our very existence. We are aware of the risk that comes with advocating for our Lands. However, it is our responsibility as Indigenous Peoples to protect our Lands and fight for them. The Land hears us take our last breath and comforts our relatives when we pass on. She is our divine and eternal mother, as she is there for us from the day we are born to the day we die. While we may lose some of our papaya trees when our Land defenders are murdered, their strong nourishment for our seedlings keeps us alive and advocating for our Lands. We must never forget the legacy of our ancestors who fought for our Lands rights. We must continue to honor their memory and fight for our Lands and rights for future generations.

Concerning Land rights, Indigenous Peoples and communities, in particular Indigenous women, must be at the forefront of any conversations. We must collectively pool our resources to amplify their Land rights movements. We must support their efforts to protect their Lands and advocate for women's Land rights. Indigenous women play a crucial role in this fight; they bring their invaluable knowledge and leadership. Data has shown us that more than 2.5 billion people live in rural and Indigenous communities worldwide that safeguard ecosystems and biodiversity for human survival.[21] Women within these communities play a key role in sustainable Land management and food security. Their contributions are vital yet often overlooked, highlighting the importance of amplifying their voices and supporting their leadership in environmental stewardship. Unfortunately, only a small portion of this Land is legally recognized, leaving these communities vulnerable to exploitative Land grabs. This lack of legal recognition undermines their ability to plan effectively and access essential government services. Therefore, we must work toward securing legal recognition and protection for these Lands to ensure the survival and well-being of Indigenous communities. As a result of securing Land rights for Indigenous women, patriarchy is challenged at its roots, and women's economic, social, and political status are fundamentally changed.[22] Of course, this requires us to unlearn and relearn many of the ways we have been taught, as patriarchy is deeply ingrained throughout global societies, including in some Indigenous communities. By empowering Indigenous women in Land governance, we pave the way for more equitable and sustainable futures for all. As climate impacts worsen, sustainable Land management becomes even more crucial. Women and girls, who are disproportionately affected by climate change, often face restricted rights to Land and resources. By ensuring their involvement and securing their Land rights, we can create resilient communities better equipped to face the challenges of a changing climate.

The pressing need to address climate change and its negative impacts cannot be overstated. The United Nations Convention to Combat Desertification (UNCCD) has recently highlighted a transformative opportunity to mitigate climate catastrophe by investing in restoring Land and ecosystems.[23] Recognizing and respecting Indigenous Peoples' Land rights, we can safeguard vulnerable landscapes and improve carbon sequestration and capture. Recognizing the critical role women play in restoring Land and ecosystems, it is crucial to prioritize their leadership in these initiatives. Indigenous women possess deep knowledge and connection to their ancestral Lands, making them invaluable stewards of biodiversity and natural resources. By supporting women in leadership roles, we can achieve better outcomes in biodiversity conservation, sustainable agriculture practices, and climate resilience. To effectively implement restoration efforts, Land rights for Indigenous communities must be secured. These

Land rights are not only crucial for maintaining cultural identities but also for preserving the integrity of vulnerable landscapes. By ensuring Indigenous communities have their rights recognized and respected, we can protect areas with high ecological value. This will minimize Land grabs and deforestation. Strengthening Indigenous Peoples' Land rights, in particular Indigenous women's Land rights, offers numerous benefits for both the environment and local communities. By safeguarding Indigenous Lands, we can prevent deforestation and natural habitat conversion to agricultural or industrial uses. This, in turn, helps maintain biodiversity, conserves critical ecosystems, and supports sustainable Land management practices. Furthermore, securing Land rights for Indigenous communities and women enables effective carbon sequestration and capture.[24] Indigenous communities have long practiced sustainable Land management practices, including rotational farming, agroforestry, and rotational grazing. These practices benefit the Land and contribute significantly to carbon sequestration. By supporting Indigenous Land rights, we can encourage these practices and increase carbon sinks, mitigating climate change.

It is unfortunate that our Indigenous communities continue to face Land conflicts that lead to displacement and violence. This is despite our efforts to protect and advocate for our Land rights. We often suffer negative impacts because of Land conflicts on our Indigenous cultures and traditions. For example, in Paraguay, the Ava Guaraní Indigenous communities once anointed newborns with kagui, a liquor made from maize.[25] These communities are increasingly disappearing in Paraguay, replaced by soybean plantations.[26] This loss represents not just a physical displacement but also a cultural erasure. The ever-expanding agricultural frontiers continue to encroach on the sacred Lands and traditions of these communities. Farmers, livestock, and even villagers are dying because of toxic agrochemicals laced throughout the Ava Guaraní territory.[27] This environmental degradation exacerbates their plight, making it nearly impossible to sustain traditional ways of life. The contamination of their

Land threatens their health and undermines their cultural and spiritual connection to their ancestral territories. Despite these dire circumstances, the United Nations Human Rights Committee has accused Paraguay of violating the Ava Guaraní rights. The United Nations Human Rights Committee cited Paraguay's failure to monitor fumigation and prevent banned pesticides in neighboring soy plantations. This negligence has resulted in severe health problems and the deterioration of Indigenous Lands and culture. Sadly, the Ava Guaraní Indigenous community had filed a criminal complaint against the regional government for failing to cease fumigation.[28] Eventually, the state launched an official investigation, but the UN reports that no meaningful progress has been made since then. The Ava Guaraní people are vulnerable, as their Land rights are constantly trampled on. Their future is uncertain due to the lack of effective action.

Indigenous communities across the Americas often lead efforts to address Land conflicts because of the large and intensive investment in agriculture corporations owned and operated privately with immense government backing. This issue is common among Indigenous communities across the Americas. Unfortunately, these agricultural industries and their expansions fuel the displacement of our people from our Lands. For example, in Guatemala, the palm oil industry expansion has spurred significant migration among Indigenous communities. Many have been forced to leave their ancestral Lands due to these corporations' aggressive acquisition and development practices. We know that this displacement journey often does not end in a fairy tale for Indigenous Peoples. When such tragedies occur to our children, we rob a part of ourselves as a community. We know that our children are our future. They are the seedlings of our culture that will sprout into our forests. However, when we lose one of our seedlings, we lose a little bit of the hope that drives who we are today as Indigenous Peoples. Jakelin Caal is a name that reminds us of all the Indigenous children we have unfortunately lost at the militarized border of the United States.[29] We lost a soul at such a young age because of the militarization of the border. She was only seven years old when she died while under the United States Border Patrol's custody. In an autopsy report, it was revealed Jakelin Caal died of streptococcal sepsis. Streptococcal sepsis is a severe infection caused by streptococcus bacteria entering the bloodstream and spreading throughout the body. Streptococcal sepsis can lead to widespread inflammation, organ failure, and, if untreated, death. Such infections are particularly dangerous for young children and require immediate medical attention. Jakelin Caal did not receive proper medical attention and died. Both the US-Mexico and Mexico-Guatemala borders are heavily militarized, severely impacting Indigenous communities by exacerbating displacement trauma and violating human rights. The criminalization of asylum seekers and climate refugees subjects them to harsh conditions and inadequate medical care.[30] This tragic and preventable incident underscores the urgent need for humane treatment and better healthcare for migrants, especially vulnerable children, at the border.

Due to Land inequality, conflicts, low wages on plantations, food insecurity, and displacement, Indigenous communities like Jakelin Caal's have been displaced. As a result of these systemic issues, many families are forced to seek better opportunities elsewhere, often in dangerous and inhumane conditions. Claudia Maquin, Jakelin Caal's mother, explained that her husband and daughter left for the United States because they could no longer plant crops. There was no Land left for them. They were driven by the hope of finding better opportunities and escaping the harsh realities of climate change and Land degradation from these monocultural agricultural corporations.

Less than twenty kilometers from Jakelin Caal's burial site, the palm oil plantation that has stolen from and severely impacted Maya Q'eqchi' communities is located.[31] The proximity of Jakelin Caal's burial site and the palm oil plantation that led to her displacement and death brings a dark shadow to the grief Indigenous Maya, in particular, the Maya Q'eqchi', are trying to heal from. Jakelin Caal's is one of the many deaths that

Indigenous Maya Q'eqchi' have experienced due to a lack of Land rights that have been jeopardized because of Land grabs and monoculture plantations.[32] Being able to see the plantation that led to the displacement of her husband and daughter, which resulted in the death of her daughter, must be hard for a mother to bear. Claudia Maquin is a strong Indigenous woman who has brought to light the terrible conditions displaced Indigenous Peoples encounter in heavily militarized borders once they encounter the borders of Mexico and the United States. Our Indigenous children are left alone, often sick. Our Indigenous parents are left to worry about their children without getting to see them or being aware of the conditions they are undergoing. It must have been hard for Jakelin's father to get the news as he was being held at a different immigration center, as family separations, which separated many Indigenous parents from their children once in the United States, continued.

Plantations such as the one located near the Maya Q'eqchi' community not only deplete the Land of its natural resources but also displace Indigenous families, forcing them into precarious situations.[33] This encroachment on their Land underscores the broader issues of environmental injustice and Indigenous territories' exploitation. Sadly, Land displacement is a major humanitarian challenge that often fuels wars. The escalating violence and intimidation by paramilitary groups and private security forces employed by these corporations exacerbate the displacement of Indigenous Peoples globally. These acts of aggression often force Indigenous families to flee their homes under threat of harm or death. This cycle of violence and displacement further entrenches Indigenous communities' vulnerability and marginalization.

Today, settler governments continue to perpetrate violence against Indigenous Peoples. While for us Indigenous Peoples this has always been evident and a part of our lived experiences, social media has brought it to the world's attention, especially when settler governments, whether the ones that govern our Lands or those from other countries, want to steal our Lands. Unfortunately, many wars and violent acts against Indigenous communities erupted when I started writing this book in 2023. Conflicts began in Gaza, the Democratic Republic of the Congo (DRC), Sudan, Haiti, and Yemen, highlighting the global scale of violence faced by Indigenous communities. The Israeli-Palestinian conflict has resulted in significant casualties and displacement of Palestinian communities in Gaza.[34] Armed groups continue to exploit natural resources in the DRC, resulting in severe human rights abuses and displacements of Indigenous Peoples.[35] Violence and instability persist in Sudan, especially in areas like Darfur, affecting numerous Indigenous communities. Political instability and gang violence have exacerbated Indigenous groups' suffering in Haiti. Yemen's protracted civil war has triggered a humanitarian crisis that adversely affects Indigenous populations. Military aggression and resource exploitation have increased in these regions, further endangering Indigenous Peoples and their cultural heritage. The fact that I saw these conflicts from behind a phone screen and witnessed the lack of support from countries like the United States to address these violent Land conflicts, in many cases genocides, made writing this book more challenging, as it focuses on climate change and displacement. The beheading of children, the death toll of loved ones, and the forced confinement of people into famines are all inhumane practices we are allowing to become normal. The normalization of violent displacement of Indigenous Peoples as a result of Land conflicts is not to be taken lightly. It is our responsibility as a global society to speak out against these injustices and advocate for the rights and protection of Indigenous communities. We must hold those responsible for perpetuating such violence accountable and work toward meaningful and equitable solutions. We can only rectify these historical and ongoing wrongs through collective action and awareness.

Today, Indigenous communities in Gaza and Congo, among others, are facing genocide, a word we are too familiar with. While many contextualize genocide as a thing of the past, Indigenous communities do not. It still happens today and severely impacts our present and future. I cannot continue this book without acknowledging what is taking place today around the world against Indigenous Peoples. This includes Palestine, Sudan, Yemen, and the Democratic Republic of the Congo, among countless countries. We, Indigenous Peoples of the Americas, know too well what genocide feels, looks like, and leaves behind. It is crucial to stand in solidarity with our brothers and sisters across the globe who face similar struggles. The ongoing conflicts have led to humanitarian crises, with countless Indigenous Peoples suffering from starvation, disease, and violence. The war in Palestine has devastated infrastructure, leaving many without access to clean water, healthcare, and necessities.[36] This dire situation has drawn international condemnation, yet the struggles continue unabated, highlighting the urgent need for global solidarity and action. It was declared in 2024 that Palestinians had the right to be protected from genocide by the International Court of Justice (ICJ), which called upon Israel to ensure such acts are prevented and that humanitarian aid would be permitted to enter the war-shattered enclave.[37] As a result of this ruling, the international community recognized the severe hardships Indigenous populations in conflict zones face. Even so, despite such declarations, these protections remain inconsistently implemented, leaving many vulnerable to further atrocities. After months of losses and despair, retribution and atrocities, the only tangible result has been compounding the immense suffering of Palestinians, with civilians bearing the brunt of power's decisions yet again. Women and children remain particularly vulnerable in this conflict zone, as evidenced by a recent report by the Independent International Commission of Inquiry on the Occupied Palestinian Territory, including East Jerusalem.[38]

Israeli illegal settlements on Palestinian Lands contribute significantly to the region's vulnerability.[39] There are estimated to be between 600,000 and 750,000 Israeli settlers living in at least 250 illegal settlements in the occupied West Bank and East Jerusalem as of 2024.[40] As a result of this ongoing expansion, limited natural resources are further stretched, making adapting to climate change even more challenging. Furthermore, Israeli illegal settlements on Palestinian Land have severe implications for climate resilience, as well as infringements on Palestinian rights. These illegal settlements consume large amounts of water, taking it away from Palestinian communities.[41] It affects Palestinians directly, and the water shortage hinders the region's ability to deal with rising temperatures and reduced precipitation.[42] Furthermore, Israeli illegal settlements have been built in ecologically sensitive areas, disrupting ecosystems and destroying natural habitats.

Besides threatening biodiversity, ecological destruction compromises the region's ability to mitigate and adapt to climate change. Israelis and Palestinians are already particularly vulnerable to climate change. Israel's and Palestine's regions are at risk of increased heat waves during the summers and limited rainfall due to desertification and declining precipitation.[43] Israel has also implemented water restrictions, which cause severe water scarcity in Palestine.[44] In recent Israeli elections, climate change has had little impact despite these vulnerabilities.[45] This is because their priority has been to continue this ongoing genocide to assert territorial control and political dominance over the region. The focus on expansion and settlement building overshadows the urgent need to address environmental sustainability and climate resilience. Consequently, the region remains trapped in a cycle of conflict and ecological degradation, exacerbating the already dire situation. The Israeli government's rhetoric on climate change and its actions are at odds because of this lack of political priority. This will lead the country to more climate vulnerability in the future. In addition, climate change has also been overshadowed by more pressing issues in Palestinian politics. Understandably, the government prioritizes short-term security concerns over long-term environmental challenges due to the ongoing conflict and political instability in this genocide. This has resulted in climate action not receiving the attention and resources it needs.

Land conflicts continue to lead to violence,[46] a violence that can no longer be hidden in plain sight. Scenes of lifeless bodies, many of them children, will haunt society for all eternity. For many years, I did not understand my father's reasoning for not telling his stories about surviving the genocide against Indigenous Maya people in Central America. When he shared these horror stories, I could only imagine them. This is because I had no idea how horrific this was to him and our communities during this time. After seeing the photos and videos of what is being done against the Palestinian people in the name of settler colonialism, my grief has resurfaced. My grief for the Palestinian Indigenous Peoples and my people who had to experience this because they fought against Land grabs. These stories fill me with sorrow and helplessness. The pain and suffering endured by both my ancestors and the Palestinian people have become a heavy burden on my heart that will never heal. It has deepened my empathy and fueled a relentless anger toward injustices committed against Indigenous communities. It has also fueled my resentment toward countries, such as the United States, that support these violent atrocities.

As Indigenous Peoples, we share many similarities in our oppression experiences. Despite the differences in our cultures and traditions, the underlying theme of dispossession and violation of our Land rights is a common bond that unites us. Our love and deep respect for our Lands will never vanish. We will continue to fight and advocate for their protection even in the face of adversity. As an Indigenous woman, I carry the weight of this history on my shoulders. My paternal communities were among those who fought and advocated for their Land rights, only to be subjected to genocide, also known as the armed conflict or civil war in Central America. The consequences of these struggles were devastating, resulting in the loss of countless lives and the destruction of our ancestral homelands. Land rights disputes are not just a professional concern for me; they are deeply personal. The loss of my ancestral Lands has affected my cultural identity but also shaped my personal experiences. The feeling of displacement and the lack of connection to my ancestral Lands has shaped my worldview and motivated me to work tirelessly for Indigenous Peoples' rights. As an advocate for Indigenous rights, my focus is on ensuring Indigenous Peoples secure their rightful ownership of their Lands. By advocating for secure Land tenure and sustainable resource management, and by fighting for the protection of Indigenous rights and interests, I aim to contribute to our communities' healing and restoration. Our love and deep respect for our Lands will never fade, and we will never cease to fight for their protection and recognition. As an Indigenous woman, my personal experiences of displacement due to Land rights disputes have fueled my passion for advocating for Indigenous rights. I hope that through our collective efforts, we can create a world where Indigenous Peoples can live with dignity and sovereignty on their ancestral Lands.

Every step in my advocacy work brings me closer to healing my community and my own wounds. It is a healing journey, both personally and collectively. The wounds I refer to are the scars left behind by historical injustices and the ongoing struggle for Land rights within my community. As I engage in advocacy work, I am constantly reminded of the resilience and strength embedded in our cultural heritage. Our ancestors have faced countless challenges and oppression, but their spirit continues to guide and inspire. Through remembering this legacy I find strength and determination in my advocacy efforts. Progress, no matter how small, is a precious and motivating milestone in my advocacy journey. Each step forward symbolizes my community's collective effort and the progress toward justice. Whether through securing a meeting with government officials, organizing community events, or raising awareness about our issues, every achievement serves as a beacon of hope. My community has supported me as an advocate. Their unwavering solidarity and collective strength inspire me to continue pushing forward, even when faced with overwhelming challenges. The support from fellow community members is the necessary boost that keeps me fighting for what is right. It is through their shared experiences and stories that I am reminded of the collective pain endured by our community. Their stories serve as a testament to the power of perseverance and the resilience of individuals. It is these stories that fuel my determination to advocate for change and work toward a future where Land rights are fully recognized and respected.

The first step toward climate mitigation and adaptation is securing Land rights.[47] Land rights are essential for ensuring access to Land and resources, and for safeguarding Indigenous communities' human rights. Additionally, securing Land rights is vital for protecting biodiversity and natural resources. Approximately 2.5 billion people depend on their Land for their livelihoods worldwide. Land not only provides sustenance but also protects ecosystems crucial for human survival. However, only a fraction of Indigenous Peoples has legal recognition of their Land. This lack of legal recognition exposes Indigenous Peoples to climate impacts and exploitation, while also exacerbating natural resources degradation. Natural resource degradation has serious consequences for global food security. As Land becomes less productive due to unsustainable practices, food scarcity becomes more likely. This, in turn, leads to increased migration and conflict as people struggle to meet their basic needs.[48] Therefore, it is crucial to recognize the importance of securing Land rights for climate mitigation and adaptation efforts. By ensuring that these Lands are officially recognized and protected, we can empower rural and Indigenous communities to manage their resources sustainably. This, in turn, can contribute to more resilient and adaptive ecosystems, ultimately benefiting Indigenous communities and global food security.

If we are like papaya trees, we play a crucial role in advocating for Land rights and we must continue to do so. Papaya trees are strong and productive plants that provide valuable resources and sustenance for communities. As papaya trees require fertile Land to flourish, Indigenous communities require secure Land rights to ensure their well-being and sustainable development. Indigenous communities' well-being and resilience depend on Land for agriculture, housing, infrastructure, and natural resources. By advocating for Land rights, we ensure that our Indigenous communities can access the Land they need. Papaya trees are known for their resilience and ability to thrive in diverse environments. Similarly, we Indigenous Peoples embody this resilience and ability to thrive. We are strong advocates and powerful tools for Lands rights advocacy. Land rights lead to increased economic empowerment and improved livelihoods.

Currently, our heavy reliance on fossil fuels for energy is what accelerates climate change impacts today.[49] This includes the continued extraction and use of fossil fuels that lead to deforestation and industrial activities that further release greenhouse gases. Climate change impacts Indigenous Peoples, especially those from the Global South, the most. Indigenous communities are often the least able to adapt to climate change, as they depend on natural resources threatened by environmental degradation. Indigenous communities are frequently excluded from decision-making processes that determine climate change impacts.[50] Papaya trees play a vital role in ecosystems, contributing to our planet's health and balance. Similarly, we play a major role in the Land rights discourse globally; by advocating for Land rights, we ensure that Indigenous communities have a say in the sustainable management of their natural resources.

4

Harvesting Our Present:
Renewable Energy

Energy is something that cannot be perceived. This is how we are taught and told to view energy when studying it at school. Despite its abstraction, energy is a fundamental concept that explains various physical phenomena. More importantly, it fuels our global societies today. Energy is the driving force behind all life and is essential for the planet's survival. It can be generated from various sources, including fossil fuels and renewable sources like wind and solar. Energy is all around us, constantly in motion and transformation.[1] It powers our homes, fuels our vehicles, and sustains biological processes necessary for our survival. From the warmth of the sun to the food we consume, energy is an integral part of our daily lives. Therefore, we cannot live without it. In biology, energy is often discussed in terms of adenosine triphosphate (ATP). This molecule provides energy for cellular processes. In chemistry, energy is considered in the form of chemical bonds and reactions that release or absorb energy. In physics, energy is defined as the capacity to do work, encompassing kinetic and potential energy. In climate science, energy is critical in understanding and addressing global warming and climate change. The earth's climate system is driven by the sun's energy, which influences weather patterns and temperatures. Even though energy has different definitions across different scientific disciplines, all disciplines disconnect the concept of energy from sociopolitical implications. By doing this, energy is differentiated from sources of injustice, environmental degradation, or climate change inequities. However, energy plays a huge role in contemporary sociopolitical landscapes, especially for those affected by climate change.

Western science removes itself from sociopolitical implications in the name of objectivity.[2] As a result, Western science is presented as acultural, apolitical, and ahistorical. Interestingly, this is how Western science presents itself given that behind every scientific concept, field of study, or framework, there is a culture, history, and sociopolitical landscape involved in its founding history or discovery. It just proves that dominant histories, cultures, and sociopolitical landscapes have become so normalized that we see them as the neutral basis for all scientific endeavors. This can further exemplify existing inequities and discrimination or further marginalize certain groups, such as Indigenous Peoples. For example, when we talk about energy in climate change discourse, we must consider how physics and climate science define it, as these definitions are integral to understanding energy systems and their impact on the environment. However, it's also necessary to recognize that these energy definitions are not devoid of cultural and historical biases. Understanding or simply being aware of these cultural and historical biases will allow further comprehension of what inequities energy as a concept already amplifies within the scientific fields. It also pushes us to look beyond the narrow boundaries that science has defined for us, given its binary and reductionary nature.[3]

We must consider when and how the concept of energy was formulated when talking about energy. The concept of energy was developed during the Industrial Revolution that England led. This was a period heavily influenced by Western scientific thought and discovery. During this time, coal began to replace wood and peat as the primary fuel source, which significantly shaped the understanding and utilization of energy. Also, during this time period, England had already depleted most of its forests due to its heavy use of wood as an energy source.[4] It had also depleted its other natural resources, thus leaving the colonial nations with little to none.[5] This is one of the primary reasons the Industrial Revolution began seeking other energy forms, such as coal. This shift in energy not only advanced industrial capabilities but also ingrained Western perspectives into the energy discourse. Oil and natural gas were used for heating and lighting throughout this era. It is important to acknowledge that European colonial countries had the mission of expanding their empires and wealth during this time. Given how Land is viewed as wealth within these capitalist ideologies, Western countries sought to colonize more Lands.[6] They had to seek Lands with untapped natural resources given how rapidly England had deforested its forests. This depletion of trees and other natural resources drove the need to explore and exploit new energy sources, further cementing the shift to coal and later to oil and natural gas. Through colonization, England pursued resources across the globe, especially in the Americas, given the rich, natural resources available there. Since our ancestors learned not to commodify nature, their Lands seemed plentiful to colonizers. As Indigenous Peoples, we are taught to see nature as an extension of ourselves, not merely as resources to be exploited. This deep connection fosters a sense of responsibility and stewardship toward the Land. Our ancestors managed and respected landscapes by building relationships with nature.

Colonizers' needs for new energy sources fueled expansionist policies, driving the exploitation of these regions, our ancestral Lands. Consequently, energy needs and colonial ambitions became deeply intertwined, shaping global geopolitics that is still relevant today. While the Industrial Revolution marked a significant shift in energy use, the preceding scientific revolution laid the groundwork for this transformation. Advances in science and mathematics during the scientific revolution led to revolutionary technologies and inventions that made extraction and utilization of energy possible. In scientific disciplines, energy continues to be taught as an abstract concept that often neglects the social implications energy systems have on society, particularly toward Indigenous communities. This abstraction overlooks the historical and ongoing impacts of energy extraction on Indigenous Lands and cultures. In addition, it illustrates the importance of not relying on an acultural, ahistorical, and apolitical lens when teaching and learning about scientific concepts like energy.

I have often been asked how Indigenous sciences can be incorporated into scientific courses and disciplines. Indigenous science is not an array of knowledge systems that can be extracted like Western science and packaged into easily accessible formats such as textbooks or lectures. Instead, one crucial approach is to unsettle the histories, politics, and cultures these fields have normalized as objective. Educators and scientists must critically examine and challenge dominant narratives that exclude Indigenous contributions. For example, the Indigenous history of energy involves understanding Indigenous communities' century-old sustainable practices and respectful relationships with the Land. As a result of this perspective, the focus shifts from simply extracting resources to valuing ecological balance and long-term stewardship. By incorporating these principles into science curricula, students gain a broader understanding of energy than they would have had otherwise. Oftentimes, I am met with the assumption that these historical contexts do not belong in science but rather in the social sciences or humanities. However, integrating these perspectives into scientific education enriches students' understanding of science as a culturally embedded practice. Additionally, it fosters an understanding of the world from a holistic perspective, which is essential for addressing complex global challenges. Especially if we are truly committed to solving climate change and achieving climate justice.

The original purpose of the energy concept was to advance capitalism and colonialism during the British Industrial Revolution, so in physics, energy continues to be emphasized in terms of work, heat, efficiency, and power. Physics plays a major role in climate science because it helps us understand the fundamental processes driving weather patterns and climate systems. However, despite the tremendous advances in energy technologies since the nineteenth century, when the energy model originated, the concept of energy as it is taught in most physics courses has not changed since. Dissecting energy definitions within physics and climate science is imperative because it allows us to learn how energy works and changes. There is a growing call to incorporate social dimensions into the education and discourse surrounding energy considering the historical, cultural, and sociopolitical background. It also helps us better understand climate change effects and the strategies we can use to reduce our carbon emissions. Scientists can better predict future climate changes and find solutions by studying the physical laws governing energy transfer, fluid dynamics, and atmospheric behavior. This knowledge is crucial for developing effective strategies to mitigate global warming impacts. Therefore, energy within physics and climate science cannot be viewed as an acultural, apolitical, or ahistorical concept.

We cannot ignore that energy aided European empires' growth by providing access to raw materials, new markets, and capital for investment.[7] All of these were founded on Indigenous Land theft and genocide. This understanding of energy reinforced colonialism by facilitating economic growth and technological advancement. Thus, the history of colonialism and the development that energy founded makes the transition to renewable energy more difficult. Acknowledging these past injustices is essential to creating a fair and equitable energy future. This involves not only transitioning to renewable sources but also addressing the grievances of Indigenous communities and the injustices hidden in the name of objectivity. The historical, cultural, and sociopolitical landscapes of energy must be highlighted. Imperial expansion and economic development relied heavily on energy resources. Energy, for example, is not an abstract concept or work capacity. It has societal implications that lead to inequities and climate change acceleration. This highlights the importance of understanding the historical context of energy production and usage. It also emphasizes the need for policies that protect the environment and ensure equitable access to energy resources. Educators, scientists, and policymakers must move beyond viewing scientific concepts in isolation, as doing so often obscures their dark histories and sociopolitical ramifications. We can never resolve inequalities or injustices without addressing the root problems but through acknowledging and addressing these complexities we can develop more inclusive and just energy policies. Understanding the sociohistorical and political context of energy can help identify more sustainable practices that are inclusive and context specific. This approach moves beyond a one-size-fits-all model, addressing unique needs and challenges in different regions and for all communities rather than perpetuating existing disparities. This is essential to dismantling systems that perpetuate inequality and environmental harm, particularly for Indigenous communities.

To begin a just transition, we must fully understand what renewable energy is and how it differs from nonrenewable energy. Both renewable and nonrenewable forms of energy have significant negative impacts on Indigenous communities. These impacts often include Land displacement, environmental degradation, and cultural disruption. Renewable energy is an energy that is continuously available and is useful to people. Renewable energy sources are often considered alternative energy sources because, in general, most industrialized countries do not rely on them as their main energy source. Instead, they primarily depend on fossil fuels such as coal, oil, and natural gas. However, as concerns about climate change and environmental sustainability grow, there is a significant push to increase the adoption of renewable energy. Renewable energy resources have lower environmental and climate impacts than fossil fuels. There are also two types of renewable energy: completely renewable and semi-renewable. Completely renewable energy sources, such as solar and wind, are naturally replenished and, essentially, inexhaustible.[8] They are not depleted by human use and can be harnessed continuously. In contrast, semi-renewable resources, such as biomass, require sustainable management practices to maintain their availability over time. Proper management involves practices, such as replanting trees and crops to replace those harvested for energy.

Contrarily, nonrenewable resources cannot regenerate or be replenished by natural processes. Nonrenewable resources have a finite supply; once they are depleted, they are gone forever. Nonrenewable resources—fossil fuels, minerals, and metals—cannot be replaced within a human lifetime. When we talk about fossil fuels, it is also imperative for community members to fully understand what they are. Fossil fuels are energy sources with high hydrocarbon content found in the earth's crust and formed from the remains of prehistoric plants and animals that lived hundreds of millions of years ago. These fuels, which include coal, oil, and natural gas, can be easily burned to release energy. As a result, they have become the dominant energy sources for industrialized societies despite their environmental impacts. Burning fossil fuels releases significant amounts of greenhouse gases, such as carbon dioxide and methane, into the atmosphere. These gases trap heat, leading to global warming and climate change.

The amount of greenhouse gases released by a society is determined by the total goods and services consumed by society. The countries of the Global South release the least amounts of greenhouse gases but face the most drastic impacts of climate change. This is because Global South countries have the lowest total value of goods and services produced by an economy, also known as the gross domestic product (GDP). With limited economic resources, these nations are less equipped to mitigate and adapt to the effects of climate change. They bear a disproportionate burden of its adverse impacts. This has led to growing poverty, hunger, and inequality in these countries. Countries place higher value on their economies, but behind their economic growth, thousands of people are losing their Lands, being forced to migrate to other countries to survive, and going hungry.

Those of us living in the Global North can enjoy luxuries denied to Indigenous Peoples in the Global South. This also includes the Indigenous

Peoples whose Lands we currently occupy, often without acknowledgment or compensation. We thrive in developed nations that continue to profit from the theft of their Lands. Considering this disparity, more awareness and reparative actions must be taken to address the legacy of colonialism. Recognition of this privilege is a step toward addressing historical injustices and supporting Indigenous rights. As people who live in the Global North, we escape the responsibility we have as humans who benefit from the immediate release of greenhouse gases into the atmosphere. We make futile decisions that make us feel better about our indirect or direct contribution to climate change. While helpful, small, symbolic actions like recycling or using reusable bags do not address the larger systemic issues. Even owning an electric car is not enough. Electric cars do not run on internal combustion engines, so they do not emit any tailpipe emissions, significantly reducing air pollution. Rather than using gasoline, electric cars rely on electric motors powered by battery packs, which are cleaner and more efficient than traditional gasoline engines. When you're driving an electric car, your carbon footprint is undoubtedly lower than that of those who drive huge gas trucks. Still, production of such cars undermines the environmentally conscious label attached to electric cars' price tags. However, while driving a hybrid vehicle may reduce personal emissions, it does not address the larger industrial and policy changes needed to combat climate change effectively. True change requires more substantial commitments to large-scale sustainable practices and policies that reduce our carbon footprint. Advocating for broader systemic transformations that include renewable energy adoption and stricter environmental regulations is imperative. We must lessen our reliance on nonrenewable resources like fossil fuels.

Furthermore, renewable energy does not mean it is not extracted from our Lands. For example, the production of batteries and the mining of necessary materials for electric vehicles, such as lithium and cobalt, can have significant environmental and social impacts. Thus, while electric vehicles are a step in the right direction, a holistic approach to sustainability must consider the entire life cycle of the vehicles and their components. It is important to understand that all energy resources have some environmental impact. Renewable energy sources, such as wind and solar, generally have a lower environmental footprint than fossil fuels. However, they are not without consequences. For example, solar panels and wind turbines can still contribute to pollution and habitat disruption. This is because both solar panels and wind turbines require large amounts of Land for installation and operation. This can lead to deforestation, and ultimately, displacement of Indigenous Peoples. This is something that many people forget to consider. Like nonrenewable energy, all types of energy must transition from upstream, midstream, and downstream to supply energy services. Starting with the extraction of the energy resource, moving through its processing and transportation, and finally delivering the energy service required for societal purposes and needs. Each stage of this transition involves various environmental impacts and challenges.

Unfortunately, when it comes to impacts on Land and natural resources, we cannot forget how this impacts Indigenous Peoples the most. Time after time, we see how climate solutions are built on the sacrifices of our Lands and Indigenous Peoples. We need a green transition away from fossil fuels, and renewable energy is a better alternative, but we cannot ignore the implications it still has on our Indigenous communities and Lands. For years, our Indigenous communities have witnessed their Lands destroyed because of the mining of rare minerals needed for renewable energy. When we drive our electric cars, we can think of the good we do for the world, but we must do diligent research to address the systematic issues. Yes, driving an electric car is an individual action addressing greenhouse emissions, but in reality, the lithium needed to build the electric car's battery was mined while sacrificing Indigenous Peoples and their Lands. The process of mining lithium also requires intensive water use, which depletes local water resources and leads to water scarcity. Lithium mining also results in the pollution of waterways, further harming both ecosystems and communities dependent on these water sources. Indigenous communities and

Land are sacrificed during this process, perpetuating the ongoing cycle of colonization that has never been truly abolished. It is critical to understand that what we view as our utopia can become someone else's dystopia.

Nevada is currently facing a dwindling water supply, and this has caused growing concern among residents. Despite these concerns, a proposed lithium mine is being constructed on Lands sacred to the Paiute-Shoshone Tribe.[9] This development has caused significant tension and division within the community. Critics argue that it infringes upon Tribal rights and undermines their cultural practices. Supporters of the lithium mine contend that it will bring economic benefits to the region and create job opportunities for community members.[10] However, opponents contend that the mine will disrupt nature's delicate balance, damage water resources, and desecrate sacred sites. They argue that the Tribe's rights and interests should be respected and protected. The proposed lithium mine has become a contentious issue in Nevada. Both sides expressed their concerns and are striving for a resolution that considers the needs of all stakeholders. The Paiute-Shoshone Tribe continues to oppose the project, but the social inequities prevent the elevation of their voices.[11] They have urged the government and mining companies to consider alternative locations or find alternative solutions that respect their cultural rights.[12] Tribal lawyers are now asking a US judge in Nevada to reconsider her earlier refusal to block digging at the proposed lithium mine near Thacker Pass.[13] Newly uncovered evidence proves this location was the sacred site of a massacre of dozens of Native Americans in 1865. This demonstrates the importance of Land to Indigenous Peoples. The Paiute-Shoshone Tribe wants to safeguard the environment for environmental, spiritual, and cultural reasons. Land is more than a physical space that can be extracted for renewable and nonrenewable energy.

When I visited the University of Nevada, Reno, on October 13, 2022, I gave a keynote presentation entitled "Indigenous Futures in an Era of Climate Displacement," in the Department of Gender, Race, and Identity. I witnessed and observed the ongoing discussions of proposed lithium mining within the university and outside firsthand. The University of Nevada, Reno has a long history with mining, even offering a mining major on its campus. As a result of this connection, the discussions were particularly poignant, highlighting the complex intersection of environmental concerns and financial interests. Even within the Tribe, discussions about supporting and opposing the lithium mine had varied opinions. This is evidence that it is critical to ensure Indigenous Peoples are not viewed as monoliths. Their perspectives and opinions on environmental and economic issues are diverse and multifaceted. While some Tribe members supported the mine dig because it would bring financial opportunities to the Tribe through revenue and jobs, this may be a short-sighted view of these complex renewable energy issues. Yes, a mine may bring economic revenue and opportunities to our people, but at what cost? Indigenous communities face degradation of sacred Lands, disruption of traditional ways of life, and potential health risks from environmental contamination. Additionally, reliance on such projects might lead to financial dependency and loss of Tribal autonomy. It is crucial to weigh long-term impacts against immediate financial benefits to ensure sustainable and respectful development.

Sadly, settler colonialism influences Indigenous communities' transition to renewable energy from fossil fuels, from Tribes in the United States to Indigenous communities and pueblos in the Global South. Energy transitions within Indigenous communities are not only a matter of capacity but embody the politics and economics of energy development in locations with long colonial histories. The history of colonialism and development affects Indigenous Peoples' current sociopolitical inequities. The United States has exerted authority over Tribal governments, compelling them to develop nonrenewable resources on their Lands as an alternative to the poverty and job insecurity their communities face. This often involves leasing or selling Land to multinational corporations that exploit these resources for financial gain. When it comes to nonrenewable energy, fossil fuel industries have been ingrained into the fabric of the

Tribes, as members depend on their employment and the Tribe's economy relies on the money generated by these industries. As a result, dependency on fossil fuels has become deeply entrenched in some Indigenous communities' culture and economy. The same settler colonial tactics are now being applied to Tribes nationwide for renewable energy projects. The loss of employment and local economic disruptions create significant social and monetary hardships that pose additional challenges during the energy transition. Traditional Land use practices and preservation of sacred sites must be considered and protected during the transition to renewable energy. The colonial legacy of Land dispossession and exploitation has left many Indigenous communities with limited resources and control over their Lands. This further complicates the transition to renewable energy and the pursuit of Indigenous sovereignty and self-determination.

There are many complexities to consider in the lithium mine and the Paiute-Shoshone Tribe situation. We must avoid condemning Indigenous community members who support the mining proposal. Especially since it is calculated that the Thacker Pass lithium mine in Humboldt County will generate more than $1 billion in investment and sales for the state each year.[14] This economic benefit could provide much-needed resources and opportunities for the local population. The community's well-being must balance these economic gains with the preservation of cultural heritage. Nevertheless, as Indigenous Peoples, we should not just agree to these renewable energy projects based on economic gain alone. We must demand what we deserve for us and future generations. Our present reality demonstrates inequities and injustices generated by these renewable energy projects. Lithium mining desecrates our Indigenous Lands. Those responsible must be held accountable. Lithium mining is sold as an essential part of the transition to renewable energy and reducing our reliance on fossil fuels. This process must be carried out with the utmost respect for Indigenous rights and Lands. By ensuring transparent negotiations and fair compensation, we can work toward a sustainable future that benefits all parties involved while protecting Indigenous Peoples and Lands.

During my visit to the University of Nevada, Reno, Daranda Hinkey's words stood out the most. Her statement, "The Thacker Pass lithium mining project will be the biggest desecration and rape of a known Native American massacre site in our area,"[15] left a lasting impression on me. Her words serve as a solemn reminder of the cultural and historical significance of our Lands. It also reminds us of the irreversible harm if Indigenous communities' voices are not respected and heard.

As I gazed across the waters of the Pyramid Lake, I felt a profound connection to the Lands and the ancestors that once walked and used these waterways for their cultural ceremonies. The serene beauty of the lake contrasted sharply with the looming threat of industrial encroachment in the name of climate solutions. Our Mother Earth is the spirit that has guided us for generations. She has allowed us to walk humbly on these grounds without viewing her as an entity that can be sold or commodified. Our Lands include our waters, air, and all living beings that share this sacred space with us. It is essential that development respects the entirety of our environment and the interconnectedness of all its elements. Protecting our resources means safeguarding our Lands for future generations' ability to thrive in harmony with nature. It is not solely Indigenous Peoples' responsibility to advocate for a more holistic and just energy transition, it is everyone's. Non-Indigenous allies play a crucial role in amplifying the voices of Indigenous communities and advocating for ethical practices in resource extraction. Through their platforms, they can raise awareness about the ecological and cultural impacts of lithium mining projects such as Thacker Pass. As allies, non-Indigenous Peoples can ensure that the transition to renewable energy is just and inclusive for everyone by standing in solidarity and demanding accountability from corporations and governments.

Colonialism has implemented that like our Lands, our bodies are also disposable. Renewable energy extraction is harsh on Indigenous Peoples and often leads to our genocide. Seeing Indigenous children's faces as they cry out from hunger, exhaustion, and despair leaves a lasting impression.

It also reminds us how our Indigenous children are disregarded under the colonizer's gaze. We have walked this earth knowing we cannot buy the sun, the air, or the clouds. We have walked this earth knowing that our suffering is for sale at the expense of others. As a climate scientist, I cannot sit at tables advocating for renewable energy as Indigenous Peoples, my people, suffer.

Cobalt is being mined off the backs of Congolese children. These children in the Democratic Republic of the Congo are being exploited, and a genocide against their people is currently being led for cobalt.[16] The imagery from the DRC, children, men, and women being forced to mine cobalt, breaks my heart. It brings forward the tears my ancestors shed as they witnessed what colonization did to our Lands and people. The DRC holds 70 percent of the world's cobalt, a mineral used to create rechargeable batteries that power our technology and electric vehicles. Children have been providing firsthand accounts of the lack of water and food and how they are being forced to sleep outside. This is all because foreign powers decided that our lives and those of Afro-Indigenous Peoples are for sale. Countries like the United States, Canada, and other countries of the Global North have aided the Congolese military in committing this genocide in the name of renewable energy. This makes me wonder whether this transition from fossil fuels to renewable energy is truly just.

In those children, I see my ancestors, grandparents, and parents, who also struggled with climate change and extractive energy projects on our Land. We were forced to flee because of the exploitation of our Land. Indigenous descendants carry the pain of displacement from climate change and genocide. As Indigenous Peoples, we are forced to be resilient, but we, too, want to enjoy our lives and thrive without inequities and barriers hindering our success. While social media has amplified these terrible conditions in the Democratic Republic of the Congo, not much media coverage has been given to what is happening there. Ironically, several climate experts claim that cobalt is the solution to efficiently electrifying the world. These climate experts report that cobalt will reduce climate change because it contributes to renewable energy production. These same experts forget to mention the severe conditions Indigenous Peoples of the Democratic Republic of the Congo and other African countries and nations are currently experiencing due to this electrification transition.[17] Only 48 percent of the cobalt found in the DRC is being mined currently. So, there is uncertainty about what will happen when they pursue mining the other 52 percent. Congo's people receive no financial gains from these mines.[18] Contrary to this, countries such as the United States, Canada, and others funnel money into this country to fund militarization for their own benefit, exacerbating conflict and instability. This external intervention prioritizes resource extraction over human rights and environmental protection, further marginalizing Indigenous communities. The international community must reevaluate its approach to ensure that renewable energy pursuits do not come at the expense of Indigenous Peoples' lives and Land.

Our humanity has failed if we are willing to sacrifice future generations to make life more comfortable for some today. We must consider the harmful effects of using both nonrenewable and renewable resources. Greenwashing is when companies publicly promote environmentally friendly practices while ignoring the human cost.[19] This deceptive marketing clouds our judgment and prevents us from demanding more ethical alternatives. True sustainability should consider both environmental and social impacts. Greenwashing is a silent genocide that has led to the displacement of over six million people since 2023 alone. As of 2020 child soldiers in the DRC have been recruited to address ongoing political instabilities due to cobalt mining.[20] I advocate for industries' histories to be highlighted within scientific frameworks and discussions, so we do not repeat harmful patterns as we see in the DRC. We cannot present the benefits of renewable energy if we forget the harm to Indigenous Peoples and Lands. When our Lands suffer, our profound roots also hurt. Our grandmother, Rigoberta Menchu, continues to remind the world that Indigenous Peoples are not simply myths of the past. She continues to utilize her media coverage to continue speaking about the injustices Indigenous Peoples continue to face in Guatemala. Children should not suffer for economic gain. Their childhoods should not be stolen for the daily comfort of other countries. Children should not be forced to take up arms to benefit their oppressors. Indigenous Peoples are what remains of what has been stolen from our ancestors; we are the warmth that comforts our communities in times of distress and need, and we are also people who should be respected and honored.

The situation in the Democratic Republic of the Congo reminds me of what my people endured, especially my father who was recruited as a child soldier due to political unrest in El Salvador. My father, like most Indigenous Maya men his age, was a child soldier in armed conflicts and civil wars in 1980s Central America. Then, our people, including our children, fought for better conditions as monoculture plantations were implemented into our Lands. Currently, similar to the people of the Democratic Republic of the Congo, our people continue to be forced to fight against the mining of rare elements needed for creating renewable energy sources. One of these examples is the Fenix Nickel Mine, located in El Estor, Izabal, the Lands of the Maya Q'eqchi' people.[21] Nickel is needed to produce the batteries used in electric vehicles and renewable energy storage. The demand for nickel drives the exploitation of Indigenous Lands and resources, perpetuating cycles of violence and environmental degradation.

The struggle for justice and autonomy remains a pressing issue for Indigenous communities. As a result, in October 2021 the Maya Q'eqchi' community of El Estor peacefully blocked the road leading to this mine for twenty days. The mine is owned by the Swiss company Solway Group and operated by its Guatemalan subsidiary, Compañía Guatemalteca de Níquel.[22] This blockade was a direct response to the environmental and social impacts caused by mining activities on their Lands.[23] The Indigenous community demanded a halt to operations until proper consultations and assessments were conducted. The Maya Q'eqchi' community argued that the mine damaged their ancestral Lands and livelihoods without providing adequate compensation or benefits. The blockade aimed to disrupt mine operations and bring attention to community demands. It garnered significant media attention and support from various environmental and human rights groups. The blockade sparked a debate about foreign companies' role in Guatemalan communities. It also inspired debate over industrial development's potential negative impacts on Indigenous Lands. During the blockade, tensions between community members and security forces escalated, leading to clashes and arrests. The same militarization techniques used during Guatemala's armed conflict, which is classified as genocide by the United Nations, were used against the protestors. Homes of the Indigenous Peoples who led this peaceful resistance movement were burned, sending the message that Indigenous rights are not respected and that the government is willing to resort to violence to protect corporate interests. This violent tactic often deployed by settler governments reinforces the distrust our communities hold against these authorities.

On April 26, 2022, Guatemala's Constitutional Court ruled that the Fenix Nickel Mine is operating illegally.[24] As a result, the Maya Q'eqchi' community has been able to validate their claims and rights. Ultimately, in 2023 the Inter-American Court of Human Rights ruled against the nickel mine, concluding that it violated the Maya Q'eqchi' community's collective rights to Land. It marks a significant victory for Indigenous rights and sets a precedent for other communities facing similar challenges. Through witnessing the effects of resisting persistently and peacefully in El Estor, Izabal, many have gained hope that justice and autonomy are achievable. The Maya Q'eqchi' community was led by the dedicated leader Rodrigo Tot, who has spent most of his life fighting for their Land rights, particularly against rare mineral mining.[25] Rodrigo Tot's unwavering commitment and leadership have been instrumental in galvanizing the community and bringing international attention to their cause. His efforts have not only helped secure legal victories but have also inspired other Indigenous communities to stand up for their rights. Still, we have to understand that standing up for our rights is not the sole solution. In many cases, like with the Paiute-Shoshone and the Democratic Republic of the Congo, it takes more than just our community efforts. We must address the power of foreign companies. They shouldn't have the ultimate say over our Indigenous Lands, but due to their economic power, they do. We have to collectively address that egregious power imbalance to make a significant and permanent change.

Unfortunately, the conquest for renewable energy sources continues to displace Indigenous Peoples, including those from the Global South. This new wave of exploitation underscores the ongoing struggle for Indigenous communities to protect their ancestral Lands and resources. The push for sustainable energy must not come at the expense of the human rights and the sovereignty of Indigenous Peoples. We are like papaya trees, resilient and steadfast, yet we are constantly under threat from those who seek to exploit our resources. The papaya tree can weather storms but still needs nurturing care; our communities need support and solidarity to thrive amid these challenges. It is not enough to resist; we must also foster alliances and seek the backing of global citizens and organizations dedicated to human rights. By standing together and advocating for sustainable practices that respect Indigenous rights, we can ensure a future where both our environment and our cultures flourish.

As I reflect on how renewable energy projects have been forced upon Indigenous Peoples, I often think of a statement made by a Liberian man I met while attending a national climate conference in Liberia. He stated, "Africa is not poor, it is overexploited." This powerful observation highlights the importance of recognizing the difference between inherent wealth and the systematic extraction of resources that leaves Indigenous communities impoverished. It serves as a reminder that true sustainability must prioritize equitable development and the protection of Indigenous Lands from exploitation. His wisdom also highlights the similarities in our struggles as Indigenous Peoples across continents and oceans. When we reflect upon what has provided nations, such as the United States and

Canada, immense wealth, we must remember what they learned from the stewardship of our ancestors. England did not have much knowledge of stewardship; neither did other colonial European nations like Spain, Portugal, and France. They did not accumulate their wealth from their own Lands but rather from colonizing our Lands. The Lands our Indigenous ancestors cultivated and cared for over centuries. Colonizers accumulate significant wealth through the exploitation of resources from our Lands and labor from our people. This left Indigenous communities with depleted resources, social disruption, and economic instability. This stark disparity underscores the long-term consequences of colonial exploitation, emphasizing the need for reparative justice and sustainable development that honors and uplifts Indigenous communities. We are exploited just like our Lands, and thus, even in the transition to a more sustainable energy system, this exploitation continues.

Papaya trees have also been displaced due to urbanization and agricultural expansion, which results in a decrease in their populations. Like papaya trees, we Indigenous Peoples continue to be uprooted by renewable energy projects. These initiatives often encroach on our ancestral Lands, disrupting our way of life and disrupting our communities. It is not surprising that at the global level, many climate solutions solely focus on environmental integrity while dismissing social integrity. This oversight leads to policies that prioritize ecological outcomes at the expense of Indigenous rights and livelihoods. Future strategies must balance both environmental and social considerations to ensure equitable and sustainable outcomes. Social integrity emphasizes maintaining community ethical and moral values, ensuring traditions and rights are respected. Environmental integrity focuses on preserving the natural environment in its interactions with humans. By integrating social integrity alongside environmental integrity, we can address issues of inequality and insecure Land rights, ensuring inclusive mitigation efforts. This dual approach not only reinforces the moral imperative of climate action but also strengthens community resilience, especially when it comes to renewable energy projects. By ensuring that both social and environmental integrity are addressed, we can mitigate the driving factors of climate displacement that Indigenous Peoples continue to experience. This holistic approach acknowledges the interconnectedness of ecological health and human rights, promoting solutions that protect both the environment and Indigenous communities. In this way, we can pave the path toward a more inclusive and just climate strategy that does not sacrifice Indigenous Peoples in the process. By recognizing and upholding Indigenous Peoples' historical rights, existing Land rights, and customary practices, we can ensure they exercise control over their Land and resources, allowing Indigenous communities to maintain their way of life while contributing to global climate solutions. Achieving true sustainability requires integrating both societal and ecological principles, fostering harmony between Indigenous communities' well-being and the health of our planet.

We should not avoid the discomfort of witnessing the pain of Indigenous Peoples. We must see and understand their hardships to move toward solutions. It is a privilege to learn about these harsh realities from a place of comfort. Acknowledging this privilege is the first step toward meaningful allyship and advocacy for Indigenous communities. By educating ourselves and taking active steps to support Indigenous rights, we can work together toward healing. However, working together does not call for the exploitation of Indigenous Peoples, their knowledge systems, and their cultures. I did not want to be displaced, but this is a byproduct of my parents' displacement. I would have rather stayed in our ancestral Lands. Even in the diaspora, our Lands call us back, but my parents were displaced because of the greed that fuels climate change, Land theft, and ongoing violence. Yet, the countries that benefit from the extraction of our Lands, especially from the renewable energy transition, are criminalizing us for the forced displacement their profit-driven endeavors are creating in our Lands and communities.

Our Lands are our community's life; they are sacred spaces where our ancestors' spirits reside and a home where cultural practices thrive. They are the foundation of our identity, providing physical and spiritual nourishment. Our Lands are the gifts we were granted by our Creator, and they hold the wisdom of our ancestors. Many Indigenous Peoples are against renewable energy projects, especially when extraction of the required minerals to fuel renewable energy comes at the expense of destroying our Lands. Protecting these Lands is not just a matter of environmental stewardship but a profound act of honoring the legacy and gifts bestowed upon us. So yes, as Indigenous Peoples, we want a renewable energy transition, but not at the expense of our Lands or our people. We demand solutions that respect our rights and recognize our contributions to sustainable practices. True progress involves collaboration that honors our sovereignty and ensures our cultures survive for future generations. However, current renewable energy projects focus on the present economic gains they can make, but not on how they destroy our Lands and create long-term environmental problems for us as a global society in the future.

In the quest for energy justice, it is crucial to consider both renewable and nonrenewable energy sources. It is no secret that renewable and nonrenewable energy continue to lead to power imbalances between energy developers who invest to make profits and stakeholders who do not have any authority to make decisions. As a result, Indigenous communities often bear the burden of environmental and social degradation without reaping benefits. Access and participation in energy decisions must be guaranteed on an equitable basis. Energy systems are mired in inequities and injustices, often reflecting broader societal and global disparities. Marginalized communities, like Indigenous communities, continue to face disproportionate exposure to pollution and limited access to clean-energy resources. Still, when we discuss renewable energy projects, the same exposure to pollution is expected to be shouldered by these communities.

I grew up in South Central Los Angeles and suffered from asthma throughout my childhood. Industries such as oil and gas, metalworking, and manufacturing facilities have polluted the air in the community in the name of profit. A battery recycling facility seven miles from downtown

Los Angeles produces lead in addition to air pollutants.[26] We, as marginalized communities, deserve to have our voices heard and our health prioritized in energy decisions. It is imperative that energy justice initiatives actively involve us in the planning and implementation of projects that impact our lives. Only then can we ensure a future where all communities benefit from clean and sustainable energy without Indigenous Peoples bearing a disproportionate amount of the environmental burden.

When I was younger, I visited my maternal Lands. The air was clean and pure. For the first time, I could breathe freely without the persistent wheezing that accompanied me in the city. Unfortunately, the air I breathe back in my maternal Lands now has drastically changed. Since 2020 over two thousand wind turbines have been installed in Mexico.[27] Wind farms contribute to air pollution through heavy machinery and materials transportation. Dust and emissions from these activities degrade the air quality. Furthermore, the once serene landscape that allowed me to escape loud city life now hums with turbine noise. This constant noise alters the natural ambiance and drowns out the birds' beautiful chirps and songs. The tranquil environment I once cherished has been replaced by industry, affecting not just humans but our plant and animal relatives. As Indigenous Peoples, when our Lands, air, and waterways get polluted, we lose an integral part of ourselves. We are the Land, and the Land is us. The sense of peace and connection I once felt has been replaced with deep sadness and loss. Indigenous Peoples grieve for their Lands and natural elements, the gifts they cherish the most because our roots are deeply embedded in the natural world. Our Lands define our identities and distinguish us as Indigenous Peoples. Our Indigenous Lands and communities are falling prey to these extractive renewable energy projects, replicating the same injustices and inequalities that came from extracting nonrenewable resources.

My beautiful Isthmus of Tehuantepec, where I hope to return once again, has now become a victim of the privatization of renewable energy projects around the globe.[28] The Isthmus of Tehuantepec, our Land, spans over four million hectares. It is also the region with the highest biodiversity in Mexico, but this rich diversity is now under threat as renewable energy projects encroach on our ecosystems. The balance and harmony that once defined our sacred Lands are being disrupted, putting both our cultural heritage and natural habitats at risk. It is these Lands that we, the Binnizá people, have stewarded along with four of our relative pueblos, the Mixe, Chontal, Ikoots, and Zoque. For generations, these Lands have been sustained and cherished by all the Indigenous communities in this region. These energy systems, despite being renewable, fracture our relationship with our Lands. Wind turbine construction has led to deforestation, soil erosion, and the displacement of our animal relatives and our communities. What is happening in my maternal Lands indicates that finite resource mining is not the only energy practice that harms our Lands. Renewable energy projects are responsible for the destruction of our Lands as well.

To get the Indigenous community's approval of the wind turbines, promises were made. Foreign corporations backed by the Mexican government guaranteed that if we allowed this wind farm to be built on our Lands, it would bring more economic prosperity and mitigate climate change. While renewable energy helps us reduce our greenhouse emissions, it is not our Indigenous communities who emit the most pollution. The opposite is true; we emit the least, yet we bear the heaviest consequences of climate change impacts and also bear the responsibility of helping countries reduce their own emissions. This irony highlights the inequity of the burdens placed on Indigenous communities in addressing the climate crisis. Indigenous communities are often the last to benefit from renewable energy sources. It is time for governments and corporations to stop looking out only for themselves. Through holding authorities accountable using global platforms and legal reforms, we can protect the rights of Indigenous Peoples.

Over four years have passed since this wind farm was built on our Land, but our communities have not seen any increase in economic prosperity.

They have only seen more injustices and negative impacts on the environment and our people. We were told that the wind farm would only take up a small amount of our Lands, but we have seen them take over much more than the estimates they shared with us stated. Although these wind turbines produce clean energy regarding greenhouse gas emissions, they are disrespectful to the Land rights of Indigenous Peoples. The corporations are taking advantage of the residents instead of paying them back their money.[29] They have also refused to offer discounted or subsidized utility rates to our local community, while energy prices are extremely high. Our resources have been exploited, and the environment has been degraded while we continue to face economic hardship. This exploitation has deepened our distrust of the corporations and the government and fueled our commitment to protect our Lands and rights. Those in power have not fulfilled their promises. Instead, we have been left to grapple with negative impacts while reaping no supposed benefits. This exploitation not only disrupts our way of life but also perpetuates the cycle of injustice against Indigenous Peoples. Our sacred spaces and agricultural areas have been destroyed, leaving us with limited resources to sustain our traditional ways of life. Also, our picturesque waterways that hug our Gulf of Tehuantepec have been overshadowed by wind farms. We used to find solace sitting near the coast, but now all we can see is the wind turbines turning and turning to generate wind power to benefit others outside our communities. When I visit my family and friends in San Mateo del Mar, Land of the Ikoots people, I can sense the disparity that has come from how wind turbines have impacted their fishing. Wind turbines have decreased fish populations, leaving many Ikoots people without a reliable source of income. This has led to continued displacement, since many have been forced to seek alternative livelihoods, often far from their ancestral Land.

The push for renewable energy has added to the complexity of Indigenous displacement due to climate change. We are not only displaced because of the impacts of climate change but also because of the so-called climate solutions imposed upon us designed without our input and implemented at our expense. Wind farms, while intended to solve climate change, contributed to poorer quality of life and displacement for Indigenous Peoples. The imposition of these projects often disregards local communities' rights and needs, leading to significant socioeconomic and cultural disruption. As Indigenous Lands are taken over for renewable energy projects, the very people who historically cared for these environments are pushed aside. They lose their homes, traditions, and livelihoods. In my homeland, wind farms have destroyed our traditional agricultural practices. Wind farms have led to the loss of fertile Land once used for growing crops and raising livestock. This has disrupted seasonal cycles and farming techniques passed down through generations, and the noise and vibrations from the turbines have affected local wildlife, which is essential to the caretaking of our agricultural systems. The number of fast-food restaurants has increased to meet the demand of foreigners who have moved to the area to work at the wind farm. So, in addition to the foreign companies that run the wind farms, foreign corporations that own fast-food chain restaurants have overrun our communities as a result of the new economic environment this wind farm has created within our Lands. The competition hinders our tias' and abuelas' ability to sell their prepared foods at the local pueblo markets due to the ability of the fast-food chain to sell cheap food quickly, without regard to culture, tradition, or health.

When reflecting, the deep sorrow we Indigenous Peoples face as we watch our papaya trees decline comes to me. This wind farm has taken over our Land, so the once-thriving orchards that sustained our people for generations now struggle to survive under towering turbines. This loss is not just environmental; it is a profound cultural and spiritual wound that we carry with us. The emotional toll is immeasurable, as these trees were not just a source of food—they were symbols of our resilience and connection to the Land. Every fallen leaf is a piece of our history slipping away. Every tree removed is a piece of our lineage lost. Every papaya trampled by machinery vanishes with our future. This transformation of our Lands leaves us grappling with a deep sense of loss and longing for the simpler, harmonious days of the past. This is why many call it green colonialism. This imposition of renewable energy projects without proper consultation and respect for Indigenous Lands and cultures mirrors the historical patterns of exploitation of colonialism. And like colonialism, this process disregards Indigenous communities' sovereignty and rights and further perpetuates silent genocides against our people. Green colonialism prioritizes profit and progress over the well-being and autonomy of Indigenous populations just as its predecessor has. It cloaks itself in the guise of environmentalism while ignoring the deep-rooted connections that Native peoples have to their Lands. This approach not only disrespects but actively harms the cultural and spiritual ethos of our communities.

With the increase of renewable energy projects in our communities in Tehuantepec, we also witnessed an increase in violence.[30] During the wind farm building, Indigenous communities experienced increased violence, and residents faced intimidation and threats from security forces hired by corporations, further escalating tensions. Since the building occurred during the COVID-19 pandemic, many of our community members who were vocal in their opposition to the wind farm were stopped at a coronavirus checkpoint. Unfortunately, fifteen people were killed in San Mateo del Mar, where this checkpoint was located. This tragic loss has profoundly affected our community. It amplifies the sorrow and anger we feel toward corporations that have disregarded our well-being and used climate solutions as an excuse to do so. There has also been an increase in sex work in the region. This has led to more violence against Indigenous women and muxes (a term from Zapotec culture in Oaxaca, Mexico, used to describe people who identify and live as a third gender) as well. This exploitation has added another layer of vulnerability to our communities, exacerbating the already dire situation.

On October 1, 2024, Claudia Sheinbaum was sworn in as Mexico's first female president,[31] taking office at a critical juncture for the nation.

As she embarks on her administration, she faces the challenging task of addressing the escalating tensions surrounding wind power projects on the Isthmus of Tehuantepec. As president, Sheinbaum must recognize the need to address these tensions and find a balance between renewable energy initiatives and our Indigenous communities' concerns. However, despite her environmental background, she is still a politician. Yes, it is progressive for Mexico to elect a woman president, but, as we Indigenous Peoples know all too well, politics often fails to prioritize our best interests. Our communities have historically been marginalized and overlooked, so we hope that this administration will genuinely consider our voices and rights in the decision-making process. Only time will tell if Claudia Sheinbaum will truly champion our causes.

Regardless of their removal, we will still have to grapple with the negative impacts that these wind farms have caused and are still causing. However, it is important to note that the negative impacts that Indigenous Peoples have faced due to renewable energy should not be used to support the continued use of fossil fuels. Instead, we need genuine, inclusive solutions that respect our rights and traditions while addressing climate change effectively. Indigenous voices must be central to environmental project planning and implementation. These energy projects should receive Indigenous communities' free, prior, and informed consent (FPIC) without misleading them or making false promises. More must be done to ensure that renewable energy projects do not harm or lead to silent genocide against our people. True sustainability can only be achieved by acknowledging and integrating Indigenous knowledge and practices and, most importantly, by respecting Indigenous rights and Lands. We do not want to hear our Indigenous children's cries as they are forced to work in harsh conditions to extract cobalt. We do not want to bury our young Indigenous women because they were murdered due to the extreme violence these renewable energy projects have introduced into our Lands. We do not want to comfort our elders as they witness the destruction of our communities and cherished environments. Yes, we strongly believe that it is necessary to find climate solutions that will help us reduce greenhouse emissions, but this responsibility should also fall heavily on the countries that emit the largest amounts of greenhouse gases. Without their commitment to action-driven solutions, our global society will continue to suffer. These nations must take the lead in implementing sustainable practices and reducing their carbon footprints. Renewable energy solutions can only take us so far, especially if the nations responsible for the most emissions do not act. Renewable energy projects should not utilize the same immoral tactics nonrenewable energy projects have utilized against Indigenous Peoples in the past. Indigenous communities must be respected and must be involved in the planning and implementation of these projects. When our Lands are impacted, our people are displaced, along with cultural and communal cohesion. This displacement not only erodes our connection to the Land but also weakens the social fabric and love that has held our Indigenous communities together for generations. It is essential to create a fair and transparent framework that prioritizes Indigenous communities and their environment. Only through genuine collaboration can we achieve a just transition to sustainable energy and truly achieve climate justice.

5

Nurturing Seedlings:
Our Youth

The gentle morning breeze floods your senses as the sun awakens alongside you. The aroma of the café de olla your aunt has generously prepared in her outdoor comal is also warming the delicious pictes you will have for breakfast. This is being home surrounded by my family and our beautiful Indigenous cultures. These are memories I treasure as a displaced Indigenous woman. These cherished scenes bring me comfort and a sense of belonging that motivates me to keep going in my professional and personal life. These sentiments are shared by many other displaced Indigenous Peoples, as the longing for home and our ancestral Lands never truly fades. For some of us, the distant memories of the times we spent in our Lands are all we have left. Alas, some of us do not have the privilege to ever go back to visit as legal immigration systems prevent us from doing so. However, this deep connection to our roots provides the strength we need to thrive in the diaspora and nurture our future generations, our seedlings.

Our memories are a reminder of the rich cultural heritage that continues to shape our Indigenous identities and guide us through life's challenges. For our children who were uprooted at a young age, these memories often fade away. These children may struggle to connect with traditions and practices that have now become unfamiliar to them, leading to loss and disconnection. They are often left confused, unable to understand why their families left their Lands. This confusion can create a profound sense of alienation, making it difficult for them to navigate their Indigenous heritage and understand their place in the world. When we are young, our understanding of displacement and the complexities of being an Indigenous person living in the diaspora is limited. While our parents and relatives instill the importance of maintaining our connections with our Lands in us, as well as the significance of our relationships with our ancestors, we often struggle to comprehend the reasons behind our physical separation from our homelands. Moreover, the absence of our extended families and communities further complicates our lack of understanding.

While I can recall how I felt as a child in a displaced family, my depth of understanding came with time and experience. Growing up in a household where my parents faced displacement from their Lands and their families, I witnessed the profound impact it had on their values and beliefs firsthand. My parents emphasized the significance of our origins and the importance of preserving our cultural identity. They taught me the history of our Lands, the spirits that inhabited them, and the role our Lands played in shaping our lives. Still, when I was young, these teachings never fully translated into a deeper understanding of the reasons behind our displacement. This is because the world outside my home confused me about identity; I was told I was Latinx by school, hospitals, and society. External labels and perceptions clashed with my parents' teachings and heritage. This dissonance made it challenging to reconcile my Indigenous roots with the identity imposed upon me by the larger world. Yet I knew I was different because the customs, stories, and spiritual practices my parents shared with me were distinctly rooted in our Indigenous heritage; they could not be truly understood under the broad umbrella of Latinx.

This stark contrast between my upbringing and societal expectations that came from stereotypes of Latinx families only deepened my sense of alienation and confusion. Unlike many of my peers, we did not have a large network of relatives living nearby. We were the only people from both our maternal and paternal lineages displaced to the United States. This absence of familial connections created a void in our lives, and we longed for connections and support from Indigenous communities. Despite the distance, my parents made sure that we carried our ancestors in our hearts. They taught us about the rich history and traditions of our Indigenous communities, reminding us of the sacrifices and struggles of our ancestors. Through stories, rituals, and oral traditions, they encouraged in us a deep appreciation for the sacrifices they made for our survival. These lessons created a sense of identity and belonging, even in the face of displacement. The adults in my life served as guardians of our culture, passing down invaluable knowledge and wisdom to us, the younger generations. They played a crucial role in maintaining our heritage and fostering resilience within the community.

It is crucial for adults to create and build safe spaces for Indigenous children living in the diaspora. These spaces should celebrate their heritage, foster community bonds, and provide support systems that nurture their growth. This is significant because we all know that when we were children, some of us did not have these safe spaces created or built for us. We often felt isolated and disconnected from our Indigenous roots. We struggled to find our place in a new world that didn't fully understand or embrace our Indigeneity. Healing our inner child means acknowledging the pain of being told we were Latinx but never truly belonging. It involves reclaiming our Indigenous pride and ensuring our children do not face the same struggles. By creating inclusive environments that honor their Indigenous roots, we help them grow with pride and confidence in their identity. We all can recall our inner child who was told they were different and excluded because no one understood our differences. I remember the sting of being labeled as "other" and the confusion of not fitting into the

Latinx category. The loneliness of those moments still lingers, shaping my willingness to ensure the next generation doesn't endure the same isolation. By sharing our stories and fostering inclusive communities, we can transform those painful memories into a source of strength and resilience for our Indigenous children in the diaspora. Our children carry the hopes, dreams, and resilience of our ancestors. It is through their growth and achievements that our Indigenous cultures continue to thrive.

The assimilation tactics used against Indigenous Peoples were harsh and led to feelings of shame and deep emotional wounds. A lot of Indigenous adults grew up being told that being Indigenous was nothing to be proud of and would lead to exclusion. These harmful narratives have made it challenging for them to embrace their Indigenous identity. For many of us, it will be the children, our seedlings, who will bring us back to our cultures. Through their growth and cultural reclamation, our children have the potential to heal these generational wounds. Our children and youth can also be our teachers. While they learn from us, we also learn and grow from them. We must create a safe space for them to find belonging and acceptance. This intergenerational healing can bridge gaps and foster a renewed sense of community and identity among displaced Indigenous Peoples. To do this, we must also provide them with the tools to navigate Indigenous culture and identity in the diaspora. Hopefully, we can incorporate teachings about Indigenous cultures from other Lands into the educational systems in the United States. By doing so, we can foster a more inclusive and accurate understanding of what it means to be Indigenous, not just from the United States but from all over the world.

Indigenous youth often play in rekindling their parents' pride in their identity. While parents may actively participate in Indigenous practices and speak their ancestral languages within their homes and communities, they may shy away from identifying themselves as Indigenous in external settings. Instead, they would focus solely on their national identities or the Latinx label imposed on them in the United States. This reality highlights the disconnect between Indigeneity and societal identity that individuals within the diaspora often navigate. This lack of public acknowledgment of Indigenous identities stems from various historical factors and the complexity of identity. It can be traced back to colonization, assimilation policies, and systemic racism. Indigenous Peoples have endured a long saga of forced displacement, cultural suppression, and erasure. This has shaped their self-perception and identity expression. Moreover, the imposition of Latinx as a label on Indigenous individuals in the United States further complicates matters. The term *Latinx*, while still widely used, encompasses a diverse range of racial and ethnic backgrounds, including European, African, and Indigenous descent. While it can be a convenient identifier for many individuals, it can also obscure and minimize the distinct Indigenous identities of those within the diaspora. It is pertinent to note that Indigenous youth's role in rekindling their parents' pride in their Indigenous identity is not solely dependent on outward acknowledgment. It is through their exploration of their Indigenous heritage, cultural practices, and language that they challenge and reshape the narratives around them. Indigenous youth are at the forefront of reclaiming their ancestral identities and advocating for Indigenous heritage visibility in public spaces. Through their activism, education, and storytelling, they challenge the erasure of Indigenous history and culture—especially those from the Global South—and urge their parents and the broader community to embrace their Indigenous roots. However, it is important to acknowledge that through social media, there is still ongoing romanticization and fetishization of Indigenous identities, cultures, and peoples. This presents a challenge as it can distort and commodify Indigenous heritage, reducing it to superficial elements that are consumed without proper context or respect. This is not to fuel the harmful narrative that all people from Latin America are Indigenous. If this were the case, our Indigenous communities and pueblos would be thriving today, not actively being displaced. It is crucial to recognize the distinct and diverse identities within Latin America and avoid generalizations that can erase the unique experiences of Indigenous Peoples. By acknowledging this complexity, we can better support and advocate for the rights and visibility of Indigenous communities and undo the harmful narratives and rhetoric that are often amplified through media.

In school, the stereotypical depiction of a Latinx household is a vibrant one that consists of multigenerational households and large family gatherings. We must acknowledge that this stereotype can overshadow Indigenous children's unique experiences within the Latinx community they are placed in. It often ignores their distinct Indigenous cultural practices, languages, and histories, leading to a lack of representation and understanding in educational settings. This can make Indigenous children feel marginalized and disconnected from their school environments. Indigenous children's reality is quite different from Latinx stereotypes. They often come from smaller families or face socioeconomic challenges that are rarely acknowledged. Their cultural heritage and traditions are overlooked, resulting in invisibility within the broader Latinx narrative. Therefore, Indigenous children grow up feeling excluded or confused about why our family dynamics are not reflected or understood. I, myself, did not have countless cousins or aunts in the United States, so I grew up in a small household of four. My family's traditions and cultural practices were distinct, and we commonly felt isolated from the larger Latinx community portrayed in the media and school curricula. This lack of representation made it challenging to connect with peers and educators who assumed a monolithic Latinx experience that did not include Indigenous Peoples. I remember feeling particularly excluded during Latinx heritage days at school, where the focus was often on popular Latinx customs that did not resonate with my Indigenous background. During these events, my attempts to share my family's unique traditions were met with confusion or disinterest. This reinforced the feeling that my cultural identity was invisible and unimportant in the eyes of my peers and teachers.

The reality of being Indigenous Oaxacan and Central American was confusing to others, even those from within our communities. During my childhood, intermarriages between Central American and Indigenous

Peoples were rare, especially given the xenophobia embedded in our communities. Sadly, Indigenous pueblos in Oaxaca rarely give refuge or support to Indigenous Central Americans due to xenophobic sentiments. Unfortunately, xenophobia was also embedded into the Oaxacan spaces I frequented from my childhood, and it still is today. This has made me resentful toward some of these spaces. We cannot truly advocate for Indigenous liberation if we do not include all Indigenous communities. My experiences have shown me that solidarity and inclusivity are essential for empowerment and recognition of diverse Indigenous identities. We need to undo the xenophobic rhetoric we are taught in and out of our households.

As a child, I knew my national identity was Mexican and Salvadoran. However, I never fit into these narrow Mexican and Salvadoran categorizations fully. My experience as an Indigenous person exposed the complexities and contradictions within such defined and articulated national identities in the diaspora. Indigenous Peoples and communities are completely erased from national identities. This is primarily due to the ongoing persecution and exclusion of Indigenous Peoples and communities. Within these national identities, Indigenous cultures, instead of being respected and celebrated, are often commodified and sold as parts of cultural histories. However, our centuries-old Indigenous cultures are rarely acknowledged for their beauty and resilience. This cycle of harm and violence is perpetuated in both of my home countries, Mexico and El Salvador, and in the diaspora, where nationalistic ideologies that oppress and marginalize Indigenous Peoples are further amplified. In the United States, these repackaged national identities often prioritize and amplify the groups and communities with the most privilege back home. It is these privileged groups who are celebrated, honored, and elevated, while Indigenous Peoples and communities continue to be overlooked and marginalized. It is crucial to recognize that these national identities, as presented in the United States, perpetuate a cycle of harm and violence against Indigenous Peoples. We have to challenge and dismantle these oppressive systems and work toward a more equitable and inclusive representation of our diverse identities. Consequently, it is not surprising that our youth are vocally denying nationalism or these national identities, especially in the diaspora, as they want to advocate for the inclusion and recognition of our beautiful Indigenous cultures. They are pushing for a more inclusive identity understanding that honors their rich and diverse heritage. By doing so, they aim to dismantle oppressive systems that marginalize Indigenous communities back in our Lands and the diaspora.

We often encounter racism within the Latinx community, despite it being labeled as a united group. This enforced label does not unify us, as oppressors will do everything to maintain their authority and further repress us. Statements such as those made by former Council President Nury Martinez in 2021 reflect our daily experiences in the diaspora and back in our homelands.[1] In such statements she claims, "I don't know where these people are from. I don't know what village they came [from], how they got here ... Tan feos [they're ugly]."[2] Her derogatory comments shed light on deeply rooted prejudice within Latinx communities against Black and Indigenous Peoples. It is imperative to note that we are not a monolith. Many individuals have benefited from Indigenous and Black oppression throughout Latin America. Martinez's comments targeted the Oaxacan community as she stated, "I see a lot of little, short dark people."[3] These remarks highlight the enduring nature of racism within the Latinx community, show the need for meaningful change, and reveal the deeply ingrained racism that permeates even marginalized people in the United States. People tend to focus solely on their own experiences of marginalization and racism in the United States, but they overlook how they perpetuate the same oppressive behaviors toward others in both their homelands and the diaspora. This selective attention hinders progress and reinforces harmful stereotypes within the community against Black and Indigenous Peoples. It was a valuable experience to witness our Indigenous youth lead the demonstrations that asked for Nury Martinez's resignation. Their activism illuminated the urgent need for solidarity and reflection within our community. It also served as an example of what our youth in the diaspora can do when mentored and supported by adults. Adults' role should be nurturing and creating safe spaces for our youth, so when they face racism in their lives, they will not internalize it. By fostering an environment of understanding and empowerment, we can help the younger generation combat prejudice and build a more inclusive community. This not only strengthens our collective unity but further paves the way for a future where all Indigenous Peoples' voices are heard and respected.

To fully advocate for Indigenous Peoples, we cannot ignore the history of colonization in Latin America, which has left a legacy of colorism and social hierarchy that persist to this day. Colonial structures established during colonial times continue to influence attitudes and behaviors within the Latinx community. In Latin America, for example, the Spanish instituted a caste system based on skin color, with lighter-skinned individuals receiving more privileges. "Mestizaje," which promoted racial mixing to whiten the population, marginalized Indigenous and African-descended people. As a result, Latin American media representation often favors lighter-skinned individuals, fostering colorist attitudes and Eurocentric practices. During my childhood, soap operas featured protagonists with lighter skin and European features in most cases. Darker-skinned actors were usually cast as subordinates or villains. The constant exposure to colorist imagery shaped my perception of beauty and worth. It showed me that society values lighter skin. It is critical to have these conversations with our Indigenous youth in the diaspora, as we come in an array of different skin colors.

By addressing the harmful impacts of colorism and promoting the beauty and worth of all skin tones, we can help dismantle these deeply ingrained biases. This is why Indigenous women like actress Yalitza Aparicio and supermodel Karen Espinosa Vega play an important role in the media. They highlight that piel morena is beautiful, challenging the long-standing Eurocentric standard of beauty. Their visibility and success not only inspire pride in Indigenous heritage but also help to break down the colorist narratives that have long dominated Latin American societies. Unfortunately, they too have faced criticism rooted in discriminatory beauty ideologies and despite their success, both Yalitza and Karen have garnered derogatory comments and discrimination based on their appearances. For example, in 2019 a racist video involving a Mexican actor Sergio Goyri circulated on social media. In the video, he makes racist remarks and belittles Yalitza's Oscar nomination. This highlights how Indigenous actors are viewed by others even back in our Lands and shows the complexities that exist because of racism. We must encourage open dialogue and create safe spaces where our Indigenous youth can discuss how Eurocentric beauty standards have harmed them and uplift their own Indigenous ideas of what is beautiful. This can help us promote inclusivity and self-acceptance among our youth and allow them to grow up unashamed of their Indigenous heritage. By fostering pride in their identity, we can prevent the cycle of shame and discrimination from continuing and proactively ensure a healthier, more confident future generation that embraces their Indigeneities.

A successful initiative is #PoderPrieto, which celebrates and amplifies the voices of dark-skinned individuals in the Latin American community. #PoderPrieto challenges colorism through social media campaigns, educational workshops, and community events. As a result of these activities, Indigenous youth can share their personal stories and experiences, thus fostering solidarity and empowerment. We should elevate more digital campaigns like this to empower Indigenous youths to find beauty within themselves. We cannot deny the impact on their self-esteem when they are constantly taught that having lighter skin is more acceptable by society. This is why we as adults should help push against the concept of Latinidad, the monolithic narrative of Latinidad that often erases the diverse experiences of Indigenous and Afro-Latinx communities, by actively challenging and dismantling colorist and racist standards that pervade our culture. Latinidad is a theory that focuses on the complex and diverse cultural, social, and political identities of Latinos and Latinas.[4] This theory explores how racial, ethnic, national, and colonial factors influence individual and collective experiences.[5] By promoting initiatives like #PoderPrieto and creating more platforms for Indigenous voices, we can contribute to a more inclusive and equitable society. Another successful and influential campaign was #LatinidadIsCancelled created by Alan Pelaez Lopez, an Afro-Indigenous, formerly undocumented poet and scholar born in Oaxaca, Mexico. Through this platform, Lopez and others have highlighted marginalized groups' unique struggles and resilience within the Indigenous and Black diaspora. Lopez's work is significant because it brings much-needed visibility to the often overlooked and marginalized Afro-Indigenous and Black Latinx communities. As a result of challenging the dominant narrative of Latinidad, Lopez contributes to a more nuanced, inclusive understanding of Latin American identity. Given the interactions our Indigenous youth, both in our Lands and in the diaspora, have now with social media, online movements like these are critical to highlight as we nourish them. Young people can connect, share stories, and find community support through these digital platforms. Supporting and amplifying these campaigns will ensure Indigenous youth feel visible, heard, and valued. For those of us with lighter skin like me, it is essential to recognize our privilege so that we can uplift these voices. Through taking part in these discussions, advocating for equitable representation, and supporting marginalized communities' initiatives, we can provide Indigenous youths essential encouragement. We must be solidaristic and active allies, ensuring our actions uplift rather than overshadow marginalized communities to achieve significant change.

Like papaya trees, when we are uprooted, we have the responsibility of nurturing our seedlings in the thorny Lands we have been displaced to. When we are transplanted from our native soil, we must adapt to survive in a new environment and face the challenges of establishing ourselves in an unfamiliar territory. Unfortunately, Indigenous Peoples have been displaced from their ancestral Lands due to various factors, such as colonization, relocation, and assimilation. Regardless, Indigenous Peoples have always found ways to preserve and celebrate their cultures. Indigenous families who are displaced often establish communities in the diaspora. This is particularly true in regions where people share similar cultural backgrounds or are located near their ancestral homelands. For example, Indigenous Peoples from Oaxaca often find communities in California, particularly in Los Angeles. Similarly, Indigenous Peoples from Central America have established communities in the DC Metropolitan area. These communities allow Indigenous individuals to find a sense of belonging and connection, regardless of their family connections. They may not be biologically connected, but it is through shared identities that individuals from various Indigenous pueblos can come together to form a common bond. This sense of community provides a support system and a platform for Indigenous cultural preservation and celebration. For Indigenous Peoples who do not have family members in the United States, these communities offer a sense of identity and belonging. They serve as a place where individuals can forge new relationships and connections with others who share their cultural roots. This sense of belonging helps to mitigate isolation experienced by displaced Indigenous individuals. These are known as transnational communities: a group of people who maintain close cultural ties across national borders. Their cultural heritages, languages, and interests bind them together and they support one another through international networks. By providing a sense of belonging and identity, these communities assist individuals in navigating the complexities of adjusting to a new environment while staying connected to their roots.

Building transnational communities helps provide a common place to hold festivities and celebrations. These events are opportunities for Indigenous Peoples to come together, share their traditions, and celebrate Indigenous cultures. These festivities not only foster a sense of unity but transmit cultural knowledge from generation to generation. Indigenous communities often form in places where they can find other Indigenous

Peoples from their Lands and pueblos. By taking advantage of social networks built by generations of previously displaced individuals, they form transnational communities that allow them to feel a sense of belonging. These networks preserve cultural practices and provide a foundation for future generations to remain connected to their heritage. They also allow us to bring a piece of our Lands and cultures to share. For example, by living near other Indigenous Peoples from our Lands, we have access to items from our homelands. When someone travels and visits their families or communities, they can bring back traditional foods, crafts, and other cultural items that help maintain and enrich the community's cultural heritage.

Encomiendas are requests made by individuals who travel back home to provide us with things we miss. These requests often involve delicious quesos (cheeses) or candies that can only be found at home. I can vividly recall the numerous encomiendas I had to fulfill, not only for my parents but also for other community members. One of the encomiendas I particularly enjoyed fulfilling was bringing back queso frijolero. This cheese, with its unique texture and flavor, was a taste of home that everyone cherished. Despite being tightly wrapped, our cheese's unique smell always permeated my luggage. This aroma was a familiar and comforting reminder of home, instantly recognizable to anyone in our community. It was a small but meaningful way to stay connected to our roots and a reminder of our shared Indigenous heritage. The aroma of queso frijolero, even from a distance, transported us back to our hometown. It served as a bridge between the past and present, evoking memories of family gatherings, delicious meals, and cherished traditions. Bringing back our queso frijolero was an endeavor that satisfied our taste buds and touched our hearts.

In addition to the regular encomiendas, individuals from our transnational communities bring items to our loved ones back in our Lands. These individuals charge a fee to our community for taking items from the United States to their families in their homeland. When I was growing up, my father had a small business selling used cars. Some community members returned to our Land in cars they acquired through my father's business and offered to bring things to our family and friends back home. This service ensured that our communities could directly send items to their loved ones back home. This was especially important for those whose family and friends lived in more rural parts of the countryside, far from post offices. The drivers would import used troquitas by tow and sell them back in our hometowns. My father went on some of these trips, and the trucks were unbelievably packed. The amount of items families sent their loved ones was remarkable. My father would meticulously organize every inch of space to ensure he could fit as many items as possible. From household goods to personal gifts, his dedication to helping our community was seen in every journey. In Mexico, the joy the family members felt when they received their items was unwavering. Many parents sent toys and gifts to their children back home. When they received their gifts from their parents, the children's happiness was beautiful. I was young, but it brought me so much joy and comfort knowing we could help, even in a small way. On these trips, we could bring more items back with us if we declared them at the border. My father always did it free of charge. He knew folks back home wanted to send something to their loved ones without having much means to do so. This generosity was a testament to his character and dedication to our community, something I grew up witnessing. Seeing the gratitude and happiness of those who received these deliveries made every trip worthwhile. Everyone knew they could repay him by preparing some nances for my mother, her favorite.

For Indigenous Peoples, food has a profound impact on our lives, connecting us to our roots, traditions, and culture. In the diaspora, maintaining our relationship with traditional foods becomes especially significant. One such example is the cherished relationship between my mother and nances. Whenever I traveled back home for the annual Feria de San Pedro festival in our pueblo, she requested nances. These small, yellow fruits held a special place in her heart. In our communities, they were always prepared by incredibly talented women to sell at this feria. They were not only her favorite fruit but also a reminder of the festivals she grew up attending as a young girl. They were one of her most cherished memories in her Land. Bringing nances back from my travels was a way of participating in the festival my mother loved. It allowed me to share a piece of her homeland with her, even if they were physically apart.

Our traditional foods are beautiful and important. They can evoke memories and transport us back to cherished moments and traditions. This is why our traditional foods hold a deeper meaning beyond mere sustenance. They are made from ingredients from the Land, making food a tangible connection to our ancestral roots. When we savor the tastes and aromas of our traditional foods, we tap into a deeper connection to our Lands. In the diaspora, where families may be far removed from their homeland, it is essential to maintain this relationship with traditional foods. By nurturing our children with these foods, we instill cultural identity, history, and a sense of belonging. They serve as a reminder of our rich Indigenous heritage and allow us to evoke happy moments. By introducing our children to traditional foods, we provide them with a nourishing diet and expose them to our homeland's rich culinary heritage. This knowledge can help them appreciate cultural diversity and cultivate a love for their ancestral roots.

Traditional foods are vital in preserving and reinforcing our Indigenous cultural identity. They carry stories, customs, and flavors passed down through generations and serve as a tangible link to our Land. By sharing and consuming these foods, we keep our culture alive. In the diaspora, it can be difficult to find traditional foods. This scarcity makes each trip back home and every package of authentic ingredients even more valuable. This lack also encourages us to seek out or even grow these foods in our new homes, ensuring that our culinary heritage continues to thrive despite the distance from our Lands. This is also a main reason why, when we are displaced, we often seek out locations where we can find other Indigenous Peoples with similar heritages. We always find ways to bring a piece of our homes or Lands with us. We open new markets that sell our ancestral ingredients from back home or start restaurants that serve traditional foods and cuisine. These efforts create community hubs that provide access to familiar ingredients and become gathering places where our Indigenous culture and traditions can be celebrated and preserved for our children and youth.

We cannot ignore that our children are often ridiculed when eating their traditional foods. This unfortunate reality is deeply rooted in the fact that our traditional foods are different from the Latin cuisine normalized and uplifted within the United States. When children compare common Western foods like pizza or french fries to the appearance of our traditional foods like mole or chapulines, our children face ridicule due to their unique culinary heritage. Western cuisine, with its emphasis on convenience and visual appeal, often emphasizes processed and standardized foods. In contrast, our traditional dishes, characterized by their rich flavors and vibrant colors, represent our ancestors' culinary traditions. Moreover, our traditional foods reflect our connection to the Land and our shared cultural heritage. They are deeply rooted in our ancestors' agricultural practices, recipes passed down through generations, and the abundance of natural resources available in our region. These ancestral connections make our traditional foods unique and irreplaceable. By placing more value on processed foods, the United States has created a disconnect between the food we consume and its origins. It makes me think of a common saying shared among my mom's friends, "En nuestro pueblos hay mucha pobreza, y aqui mucha harmburgesa." The saying highlights that while poverty is abundant in many of our ancestral communities, processed foods are easily accessible and abundant here. This disconnect has devastated our health, culture, and economy, especially in the diaspora.

We already face health disparities in the diaspora due to our communities facing extreme pollution, but our diets also play a role. The shift from traditional, nutrient-rich foods to heavily processed and convenience-based options has exacerbated chronic health conditions. For instance, our community experiences higher rates of diabetes, obesity, and heart disease than the general population. This has been exacerbated by limited access to fresh, healthy foods and the prevalence of food deserts in the areas where many of us live. The stress of cultural assimilation and the loss of traditional dietary practices compound these health issues, making it even more challenging to maintain our well-being. This is why it is crucial to address cultural bias around traditional foods and our children to preserve our ancestral practices and strengthen our communities.

Nurturing our seedlings, the younger generations, in our newly built homes can be challenging but rewarding. It requires patience, dedication, and flexibility to adapt to new circumstances. Like papaya trees, we must use our extended roots beyond our Lands to grow and flourish in our new environments, and we must provide the necessary care and attention for our seedlings to thrive. This may include providing a suitable environment, such as adequate sunlight (our Indigenous languages), water (our Indigenous traditions), and soil nutrients (our traditional foods). We also need to protect them from pests and diseases such as settler colonialism, exclusion, and assimilation that threaten their survival. Nurturing our seedlings takes time and involves establishing connections with others in our transnational communities. In the same way papaya trees rely on their support networks, we, too, can benefit from building relationships. We can find solace in those around us who offer support and guidance. As we each have our own goals and dreams, each seedling has its own needs and requirements. We can overcome challenges and thrive in our new homes by remaining flexible and adaptable.

Despite being away from our Lands, our Indigenous languages can provide the sunlight needed to flourish in the diaspora. This means we must focus on language preservation and teaching so that our Indigenous languages are passed down to future generations. Through integrating language learning into everyday activities, we create a strong foundation for cultural identity. Storytelling, traditional songs, and communal gatherings keep our linguistic heritage alive. Many adults, bearing the scars of historical trauma and forced assimilation, may feel disconnected from their mother tongues and hesitate to teach them to their children. This reluctance can stem from a fear of their children facing similar marginalization or a loss of fluency over generations. Language loss is not only happening in the diaspora but also back in our Lands. Colonization, globalization, and the dominance of languages such as Spanish and English have contributed to the erosion of our linguistic heritage. The Instituto Nacional de Lenguas Indígenas (INALI) has found data that many Indigenous languages, including the Zapotec language, are gradually losing fluent speakers, especially among younger generations. While multiple factors play a role in this decrease in fluent speakers, displacement and immigration are the number one reasons. This makes sense because, in the diaspora, our community members have to navigate another colonial language, English. Most bilingual and translation services are offered in Spanish, forcing our communities to enhance their Spanish and focus on teaching this to their children over their native languages. This dual pressure makes it even more challenging to prioritize and maintain Indigenous language instruction within the family and community settings.

My mother spent countless hours teaching me Spanish. I did not understand why she was so motivated to ensure I could comprehend Spanish at a high reading level. Every weekend, we read Spanish books, and she helped me pronounce the words. Our time commitment to learning Spanish allowed me to move into a higher-level class when we transitioned into the bilingual portion of the class. I remember going to the second grade for reading time, as I was too advanced for the first-grade-Spanish reading level. While this gave me an academic advantage, it also meant that the time spent on Spanish took away from learning and using our Indigenous language. As a result, my proficiency in our Indigenous language remained limited, and I struggled to communicate with older family members who were fluent. This experience is common among many members of our community, which contributes to a generational gap in language skills. Over time, this gap widens and threatens to erase our Indigenous languages. However, my experience and my mother's determination to teach me Spanish stemmed from the challenges she faced growing up without understanding Spanish. She recounted stories of feeling isolated and discriminated against because she couldn't communicate effectively in the dominant language in her homeland. Throughout history, Indigenous Peoples have been portrayed in the media, especially in telenovelas. Actors who play Indigenous roles often speak Spanish with a heavy accent to ridicule and caricature Indigenous Peoples' language. This portrayal perpetuates stereotypes and reinforces negative perceptions, further marginalizing Indigenous communities and making us ashamed of our Indigenous languages. Media representations contribute to Indigenous languages' stigma and discourage use among younger generations.

My father often felt shame because he did not know how to read or write in Spanish. This limited his opportunities and interactions in the diaspora. This sense of inadequateness reminded us of our community's barriers. His experiences reinforced my mother's determination to ensure I would not face the same struggles. Yet, it highlighted the complex balance between embracing a dominant language and preserving our Indigenous heritage. I recall the many times my father practiced writing Spanish alongside me, with my mother as our teacher. These shared moments of learning created a supportive bond in our family. It also allowed me to understand my parents' connection to the marginalization they faced back home speaking their Indigenous languages. These memories allowed me to understand why my parents emphasized that I learn Spanish. They wanted to shield me from the discrimination they experienced and provide me with better opportunities. Through these lessons, I became proficient in Spanish and gained a deeper appreciation for my Indigenous roots. Sadly, these experiences are very similar in our Indigenous communities in the diaspora as well as at home. Despite the hardships, these experiences highlight the importance of preserving and valuing our Indigenous languages and cultures.

As an educator in Little Rock, Arkansas, I had a student in my class who was an Indigenous person from Mexico. He was not proficient in Spanish. Despite the teacher's efforts to provide him with translated materials and assistance, he struggled to understand the content because it was in Spanish. It became evident that his primary language was an Indigenous language not commonly recognized by the school system, especially in bilingual education. This revelation led me to understand his unique challenges. It also made me ponder how other Indigenous students who came to the United States as unaccompanied minors were doing, especially at school. The student's family speaks Tzotzil Maya. Tzotzil Maya is a language spoken primarily in Chiapas, Mexico, and belongs to the Tzotzil subgroup of Mayan languages. While this language is native to his family, it was not recognized by the school system, leaving him at a disadvantage when learning Spanish. Initially, teachers were puzzled by the student's inability to comprehend the subject matter even when translated into Spanish. They assumed he had a learning disability that hindered his ability to understand the material in Spanish as well. However, the revelation that his family spoke Tzotzil Maya shed light on the student's complexity. Teachers realized that his family was actively teaching him Spanish, most likely in response to his request for school assignments and homework in Spanish.

Since 2021 to 2022, 926 unaccompanied minors who migrated into the United States have been released, according to the Administration for Children and Families. The number of unaccompanied children crossing the US and Mexico border has increased notably. This highlights the growing need for resources and tailored support for these young migrants, especially Indigenous children. This surge underscores the importance of recognizing and accommodating diverse linguistic backgrounds of students like my Tzotzil Maya–speaking student. Addressing these unique educational needs is crucial for successful integration and academic progress. Educators witness firsthand the challenges these children face when adapting to an unfamiliar environment. However, in states like Arkansas, where immigration is addressed differently than in more liberal or democratic states, there are often no comprehensive support systems for these children. This can lead to significant challenges in their education and overall well-being. Educators, who understand their situation, must step in to bridge the gap, providing academic and emotional support. This example is far too common, especially when it comes to Indigenous children who are displaced before they fully grasp the Spanish colonial language. These children face layers of linguistic and cultural barriers, making their educational journey even more challenging. It also adds to the challenges parents and adults face when teaching their children Indigenous languages. Indigenous languages are a significant part of our culture, and we must ensure that children have access. However, if we continue to promote only Spanish in education systems, it puts our Indigenous children who are displaced, our seedlings, at a disadvantage. This narrow focus neglects the rich cultural heritage and linguistic diversity these children bring. By failing to support their native languages, we risk alienating them and hindering their academic and personal growth. This is why we must interrogate the Latinx identity and label. As it stands, this label comes with the Spanish language attached to it. This is the same situation our Indigenous children face back in Mexico, as school and education are offered in Spanish alone. This systemic oversight disregards their Native languages and cultures, further marginalizing these communities. Therefore, it is crucial to advocate for inclusive educational policies that honor and integrate Indigenous languages in the US and Mexico to support our Indigenous communities in language revitalization.

Fortunately, there are language revitalization initiatives taking place both in the diaspora and back in our Land. These efforts help to celebrate, pass down, and preserve Indigenous languages and cultures for future generations. In 2018 Oscar Méndez Espinosa published the first Zapotec dictionary, comprising five volumes and sixty thousand words. In this monumental publication, the basic vocabulary of all Zapotec-speaking communities and its sixty-two variants is collected in Oaxaca. A dictionary helps learners by serving as a guide for correct spelling and pronunciation, making it an indispensable tool for language acquisition. It provides comprehensive definitions and usage examples, aiding in the understanding and proper use of the Zapotec words. This facilitates more accurate communication and deeper engagement with the language. The inclusion of several language variants allows for several Zapotec pueblos to be included, as each speaks a different variant according to where they live. This is significant because it also acknowledges linguistic diversity within the Zapotec community. This approach ensures that the dictionary is a comprehensive resource, respecting the cultural and linguistic nuances of each variant. By doing so, it fosters a deeper understanding and appreciation of the rich tapestry of the Zapotec languages. Universities and academic institutions also play a major role in language revitalization and learning.

Universities and academic institutions also play a pivotal role in language revitalization. They offer specialized programs, courses, and research opportunities focused on Indigenous languages, which train the next generation of linguists and educators. These efforts support language preservation but facilitate and promote active use in educational and community settings. These efforts involve collaboration with Native speakers and community leaders. These academic efforts provide resources and research and foster the next generation of linguists and cultural advocates. For instance, the Metropolitan State University of Denver offers immersive programs that support the revitalization of the endangered Maya Ch'orti' language. The University of New Mexico's Latin American and Iberian Institute offers courses in Quechua, Nahuatl, and Yucatec Maya. It is through the provision of such courses that the institute contributes significantly to the preservation and revitalization of Indigenous languages. While these efforts are important and helpful for the revitalization of our Indigenous languages, it is also crucial to ensure that Indigenous youth in the diaspora are afforded these opportunities. Providing access to university programs and resources can empower them to reconnect with their heritage and contribute to ancestral language preservation. By including the diaspora in these initiatives, we encourage a more robust and inclusive approach to language revitalization. This is why it is critical to interrogate these academic institutions, as Indigenous representation is relatively low. Are these programs accessible to all, and do they address the unique challenges youth face living away from their ancestral Lands? These questions are essential in evaluating the true impact of language revitalization programs. Accessibility and inclusivity must be prioritized to ensure that all Indigenous youth can participate and benefit regardless of their geographic location. Addressing these unique challenges will enable a more comprehensive and effective approach to preserving and revitalizing Indigenous languages.

Additionally, Indigenous communities are using culturally relevant methods. In Alta Verapaz, Guatemala, the Q'eqchi' Maya community is leveraging community-based radio stations to revitalize their language.[6] These radio stations broadcast in Q'eqchi', providing a platform for storytelling, education, and cultural exchange.[7] This initiative promotes language use and strengthens community bonds and cultural identity. Radio is an accessible medium that can reach a wide audience, including those in remote areas. It provides an engaging way to deliver educational content, making learning more interactive and culturally relevant. Radio fosters a sense of community and belonging, as listeners can tune in to hear familiar voices and stories in their native language. The example above, as many others, illustrates the indigenuity of our communities and how they use them to revitalize our language.[8] Our Indigenous indigenuity is evident in the creative and resourceful methods employed to ensure the survival and thriving of our languages. By merging traditional knowledge with modern technology, communities are finding innovative ways to keep their cultural heritage alive. This blend of old and new demonstrates a profound resilience and adaptability that is essential for the preservation of our linguistic and cultural identities.

Technology plays a pivotal role in Indigenous languages revitalization. Digital platforms, such as mobile apps and online language courses, provide convenient and accessible ways for individuals to learn and practice these languages. This is regardless of their geographical location. Furthermore, social media and virtual communities offer spaces for speakers to connect, share resources, and foster solidarity in their language preservation efforts. In addition to radio stations, technology is being applied through the creation of language learning apps and interactive websites tailored to specific Indigenous languages. These digital tools offer immersive experiences with audio, video, and gamified learning modules.

In 2023 the Mexican National Institute of Indigenous Peoples (INPI) announced an upcoming national university, the University of Indigenous Languages of México (ULIM). The university, which opened later that year, specializes in Indigenous language courses and degrees.[9] This initiative aims to provide formal education and training in Indigenous languages, ensuring their preservation and passing to future generations. The ULIM offers three degrees: Teaching of Indigenous Languages, Interpretation and Translation of Indigenous Languages, and Indigenous Intercultural Communication. This diverse range of programs equips students with the skills necessary to become educators, translators, and cultural ambassadors for their communities. By providing formalized education in these areas, ULIM plays a crucial role in the revitalization and preservation of Indigenous languages and cultural practices. ULIM also provides opportunities to build relationships between Indigenous and non-Indigenous communities, creating a strong foundation for understanding and collaboration. This benefits all members of the community, both culturally and socially. Programs of this type also highlight Indigenous languages' cultural significance in the community, fostering greater appreciation within our communities. It helps us understand that Indigenous languages are important and heal from the scars that speaking our Indigenous languages left on us because of the oppression we face for doing so. Additionally, they provide economic opportunities for fluent speakers of Indigenous languages, allowing them to use their linguistic skills in various fields. This dual impact strengthens both the cultural and economic fabric of Indigenous communities.

Within our Indigenous languages, art is strongly integrated. This demonstrates how innovative our Indigenous languages are, as they beautifully weave together visual and verbal expression. Unfortunately, it also underscores the importance of revitalizing them, as these unique cultural elements risk being lost. Preserving and nurturing these languages are crucial for maintaining the rich heritage they embody for our Indigenous cultures. In *Distant Readings of Disciplinarity*, Miller documents how our language images are sonic, pictorial, gestural, and alphabetic.[10] This multifaceted nature of language showcases Indigenous communication's rich and expressive capabilities. While Western languages often focus primarily on alphabetic and verbal communication, they lack holistic integration of multiple modes of expression. This divergence highlights Indigenous languages' unique and comprehensive nature. By revitalizing these languages, we preserve their multifaceted and profound storytelling methods and cultural expression for future generations. Our seedlings need to be nourished by our Indigenous languages.

It is also not a surprise that our languages are woven into our artwork, from clothing to our artesanías. Our language breathes life into every creation, from intricate textile patterns to pottery with symbolic motifs. This interconnectedness highlights the deep cultural significance embedded in our artistic traditions. The gorgeous huipiles tell a story about the person who embroidered it. They serve as an expression of their nature and are often their deepest thoughts. The wood carvings are made while navigating complex mathematical language expressions that cannot be fully comprehended by Western scientists, even today. Our handcrafted beaded jewelry highlights familial ties, social status, and spiritual beliefs. Each bead and pattern is intricately linked to our ancestral knowledge, medicine, and traditions. This jewelry not only serves as an adornment but also as a living testament to our cultural identity and history as Indigenous

Peoples. In our transnational communities, we have stores or families who solely focus on selling artesanías. Indigenous youth should be encouraged to learn about and engage with these traditional art forms. They can thus contribute to preserving and continuing our heritage's cultural practices. By keeping our cultural identity vibrant and alive, this involvement ensures that knowledge and skills are passed down through the generations, those in our Lands and those in the diaspora.

Back in our Lands, I was taught to bead and embroider by my relatives. I always remember them asking me why I wanted to explore these art forms. This frequently puzzled me. It was something I was proud of. I valued that my aunts and uncles could create beautiful art from their hands. As time passed, I learned that they interrogated why I wanted to learn due to society's lack of appreciation for our artesanías. Thus, this deeply ingrained sentiment has prevented the skilled generation from passing their artforms to the younger generation. We must actively prevent this erasure as our artesanías carry our stories. We express our cultures on canvases made of wood, velvet, or other mediums. One way to preserve these traditional art forms is to create community workshops where elders can teach the younger generation. Integrating these arts into school curricula can help to instill pride and appreciation from an early age. Lastly, using social media platforms to showcase and sell these crafts can increase their visibility and value in modern society. Social media can help us attract thousands of followers and connect artisans with a global audience. However, it is critical that we also protect our artesanías and art from exploitation. This applies to those who want to benefit from our art without giving our artesanos proper credit. Racism is deeply embedded in the exploitation of Indigenous Peoples and art forms.

Like many displaced plant relatives seen as invasive species, we must actively find our place in the diaspora. We must nourish and heal our inner child by embracing our heritage and addressing past traumas. We must foster a sense of belonging and resilience in the diaspora through sharing the fundamental teachings passed from elders and parents. Despite our roots being stretched from our Lands to new soil, the nourishment we give our seedlings in the diaspora also supports the seedlings planted back home. Strength and solidarity are created by shared traditions and wisdom that transcend geographical boundaries. Despite their distances, both communities can thrive and grow because they are interconnected through our Indigenous roots. We must walk a fine line, as we do not want to negate the experiences of the Indigenous Peoples of the Lands we are displaced from. We must build respectful relationships with the original inhabitants and acknowledge their sovereignty and cultural heritage. My grandmother's teachings come to mind. We are unwelcome guests until we build genuine relationships with the Indigenous Peoples of these Lands. She always said respect and humility are the first steps toward genuine connection and understanding. Through listening to and learning from the original stewards of the Land, we can create a path forward. Educating our youth in these values ensures that our culture will thrive. Through our nurturing, they develop a strong sense of identity and connection to their heritage. The youth will become ambassadors of our shared history, bridging the gap between our ancestral roots and the new communities we create.

6

Protecting Our Roots:
Climate Justice

I always say that we all have our radical origin story. That moment was when we decided we were not okay with the status quo and wanted change for ourselves and our communities. When I witnessed how Mexican soldiers and policemen mistreated my father at checkpoints, it made me perceive society differently. On our road trips to Mexico, I witnessed several instances of oppression toward him. As a little girl, I was confused, but I knew he was receiving different treatment and wondered why. I felt sadness and frustration watching someone I cared about being treated poorly. It was difficult to understand why such inequality existed, and it left me feeling helpless. These experiences shaped my awareness of social injustice and helped me understand my identity.

I remember the countless mordidas he had to give them to stop the harassment and mistreatment he received. Mordidas are bribes expected by Mexican soldiers and policemen, especially from Central Americans. This highlights corruption in Mexico and how soldiers and police officers oppress Central Americans. Whenever we reached a checkpoint and they saw my dad's Salvadoran passport, they demanded mordidas. My father had permanent resident status in the United States, but the checkpoint authorities always asked him what his country of origin was. Only cars with Central Americans had this problem. My dad is friendly and approachable, so he engaged with other travelers who were also asked to step to the side. Many were also from Central America and were returning to their homelands to see loved ones. Others drove back home to sell their cars and deliver goods and gifts from displaced family members. At these checkpoints, Mexican policemen and soldiers understood this, so they knew they could easily get mordidas from the travelers. I remember the mix of fear and defiance in my voice as I asked them why my father had to pay. The officers simply smirked, dismissing my question without a second thought, as if it were just an insignificant annoyance. That moment ignited a fire within me. Since then, I have been determined to stand against such injustices and fight for a more equitable world.

Experiencing injustice and mistreatment at a young age leaves a deep and lasting impression. This shapes our understanding of the world and instills a sense of caution and distrust toward authority figures. We form our identities and navigate society as adults based on these early encounters with discrimination. My experiences radicalized me, causing me to become more conscious of global inequalities and taught me the importance of standing up against injustice and using my voice to bring about change. It inspired me to become more involved in activism and advocacy, pushing me to fight for justice and equality. Young people are often led to believe that our voices lack power or significance, but the impact of our experiences with inequality never truly fades from our minds. As we grow older, we realize that our voices can be powerful tools for change and that speaking out can inspire others to join in the fight against injustice. Being Central American in Mexico should not be a crime. Being Indigenous in Mexico should not be a crime. Still, the harsh realities we continue to face are that simply existing as a Central American or Indigenous person in Mexico often feels like an act of defiance against a system that views our identities as criminals. The pervasive prejudice and discrimination we encounter highlight the urgent need for systemic change. Despite these challenges, we must continue to advocate for our rights and demand recognition and respect for our communities.

I remember the countless times my dad got asked if he was an Indio, a question laced with disdain and prejudice. Throughout each encounter, he and, by extension our family, were reminded of the ingrained societal biases, both here in the United States and my maternal homeland, Mexico. My experiences strengthened my resolve to challenge these stereotypes and fight for a more inclusive and equitable society.

As a result of these experiences, I am convinced that every aspect of Indigenous culture and identity is political and radical. From our languages and traditions to our Land rights and self-determination, each element challenges the status quo and demands acknowledgment. This perspective shapes my approach to advocacy and activism, as I strive to honor my heritage while pushing for meaningful change. No matter our nationality, if we are Indigenous, our existence, survival, and resistance are intrinsically political. Our very presence defies centuries of colonization and marginalization, asserting our right to thrive on our terms. This shared struggle unites us across borders, reinforcing the urgency of our collective fight for justice and recognition. Our presence alone challenges historical and ongoing injustices in our fight for rights and recognition for our communities. By asserting our identities and advocating for our rights, we engage in a political act that seeks to reshape the systems that have marginalized us since colonialism. Through our activism and cultural expression, we are reclaiming our narratives and paving the way for a more equitable future.

In many countries, gaining official recognition as Indigenous Peoples involves navigating complex legal and bureaucratic systems. This requires engaging in government policies that have historically ignored or undermined Indigenous identities. As a result, recognition becomes a form of political activism that challenges existing power structures. Therefore, being Indigenous means being radical. A radical identity embraces and celebrates our unique cultural heritage and traditions while actively resisting and destroying oppression. It means unapologetically asserting our rights to Land, language, and self-determination, while refusing to conform to the dominant narratives that erase our existence. This radical identity empowers us to forge alliances, build solidarity with other marginalized groups, and envision a future where diversity and justice are at the forefront of societal values.

Today, being radical is perceived as extremist.[1] Radical thinking, however, often involves challenging the status quo and advocating for necessary social change. By labeling radicals as extremists, society can undermine their efforts and obscure the positive aspects of their movements. This tactic creates fear and division, preventing meaningful progress from occurring. Embracing radical ideas in the context of the environmental crisis is crucial. This is controversial because it challenges entrenched interests and demands systemic change. We must recognize the urgency of this moment and work together to create a more equitable and sustainable future. It is through collective action that we create meaningful and lasting change.

As climate change intensifies, the number of climate refugees increases rapidly, displacing countless individuals from their homes. This phenomenon underscores the urgent need for radical solutions that address environmental degradation's root causes. Indigenous Peoples have long been on the frontlines of climate change, experiencing its impacts firsthand. Our traditional knowledge and Indigenous science coupled with our sustainable practices offer invaluable insights into how we might live in harmony with the earth.

In 2020 a viral social media statement captured the stark reality of our times: "You are significantly closer to being a climate refugee than a billionaire."[2] This poignant observation underscores our vulnerability in the wake of climate change, emphasizing the need for immediate and transformative action. Climate change impacts are being felt all over the world, from the melting of polar ice caps to the displacement of entire communities. We are seeing an increase in climate refugees who are forced to leave their Lands and homes behind because they are no longer inhabitable. When it comes to climate justice, we ask, "What needs to happen to provide climate justice for climate refugees? They have already been violently uprooted from their Lands, so can we help them achieve climate justice?" Ironically, while many are encouraged to chase wealth and success, climate change means more people face displacement. The focus should shift from individual wealth accumulation to collective action and support for those affected by climate crises. Only by addressing the root causes of climate change can we hope to achieve true climate justice for refugees and everyone, especially Indigenous Peoples. We cannot be oblivious to fossil fuels because we all chase this dream of becoming wealthy and rich. One day we, too, will face climate change impacts that lead to displacement. Sadly, human nature often dictates that people only take action when personally affected. This tendency to react rather than proactively address issues means that many people do not prioritize climate justice until the repercussions land on their doorstep.

It is not always a good thing to strive for wealth and riches, especially when data has shown that the richest 1 percent emits as much pollution as two-thirds of humanity.[3] This stark inequality in emissions highlights the disproportionate impact that a small fraction of the population has on the environment. The data also illustrates how wealth and riches are built by sacrificing others for our own needs. The 1 percent is often more focused on maintaining their luxurious lifestyles, even if it means contributing to environmental degradation. Instead of taking action to reduce their carbon footprints, they continue to engage in activities that accelerate climate change. This highlights a significant disparity in responsibility and accountability between the world's wealthiest individuals and the broader population. Society must demand more transparency and accountability from these individuals, particularly regarding their environmental impact. We can encourage more sustainable practices by holding the 1 percent responsible for their actions. In 2023 public data revealed that several high-profile celebrities were among the top contributors to private jet carbon emissions.[4] This information sparked widespread public outcry, highlighting the need for these figures to lead by example in adopting more sustainable travel practices.[5] The revelations served as a powerful reminder of the environmental impact individual choices can have, urging a reevaluation of lifestyle habits among the elite. While it is important to respect individuals' privacy, the excessive use of private jets by the elite poses a serious threat to efforts to combat climate change. These trips contribute disproportionately to carbon emissions, undermining global sustainability goals. Encouraging these individuals to reconsider their travel habits is essential for fostering a more environmentally conscious society.

Addressing this imbalance is crucial to achieving true climate justice and sustainability. This means that a tiny segment of the global population is contributing massively to the climate crisis, while the vast majority bear the brunt of its impacts. To achieve climate justice, we must hold these high emitters accountable and implement policies that reduce their disproportionate carbon footprints. Doing so will pave the way for a more equitable distribution of the burden and benefits of addressing climate change. As calculated in 2019, the richest 1 percent, approximately seventy-seven million people, emit half of global emissions.[6] That should not be the case. This staggering disparity reveals the extent to which the wealthiest individuals contribute to the climate crisis. It highlights the urgent need for policies which curtail excessive emissions from the top 1 percent,[7] but promote sustainable practices across all sectors. According to data, the poorest 99 percent of global citizens would need 1,500 years to produce as much carbon as billionaires.[8] This data illustrates the vast chasm between the carbon footprints of the wealthiest and the rest of the population. Such an extreme disparity underscores the pressing need for systemic changes that target the ultrarich's excessive emissions. By addressing these imbalances, we can move toward a future where environmental responsibility is more evenly shared.

At the time of writing, it feels as if we live in a parallel universe. We are witnessing how wars are exacerbating the human impact on our planet. Wars produce high amounts of greenhouse gases that contribute significantly to anthropogenic climate change. They also result in pollution that contaminates air, water, and soil, disrupting ecosystems and harming biodiversity. Additionally, wars lead to resource depletion as natural resources are exploited and destroyed, further straining our planet's already limited reserves. The answer to whether wars are necessary is always no. Wars often arise from perceived threats that the leaders of powerful nations perceive from marginalized communities or countries. These conflicts are frequently rooted in a desire to maintain control or power, rather than addressing the underlying issues that might lead to a more peaceful resolution. This cycle of aggression not only endangers the planet but also perpetuates the marginalization of those already vulnerable.

Oftentimes, climate refugees have experienced war within a short time frame from when they were forced to flee their Lands. Their Lands are already vulnerable to climate change impacts, so wars end up accelerating these impacts and make their Lands uninhabitable. The combination of environmental degradation and conflict leaves these communities with few options but to seek refuge elsewhere. Pressure on neighboring regions and countries increases, creating further challenges in providing adequate resources and support. Yet, as global citizens, we cannot turn our phones off because we do not want to face reality. This is a reality Indigenous Peoples know and have experienced before. What my people experienced, including my father and uncles, is now happening today, in front of our own eyes. Their stories of displacement and survival echo the current plight of countless individuals facing the same hardships that are amplified a thousand times because of the current wars. Past lessons go unheeded, as history repeats itself with devastating consequences for upcoming generations.

As a result of three consecutive years of drought coupled with multiple conflicts, livelihoods have been decimated across Ethiopia, leaving communities unable to sustain themselves.[9] As a result, El Niño–induced flooding could worsen the already dire situation, displacing even more people and destroying what resources are left. This confluence of environmental and human-made disasters shows the urgent need for comprehensive solutions that address climate change and conflict. In November 2022 there was a ceasefire between the Ethiopian government and the Tigray People's Liberation Front (TPLF).[10] Despite this, ongoing conflicts in the central Oromia region and the Amhara region in the northwest threaten to unravel any peace efforts.[11] Following the coup in July 2023, Niger has experienced political tension with its neighboring countries, with international security assistance withdrawn.[12] This has exacerbated existing vulnerabilities, leaving the region more susceptible to internal and external threats.[13] The absence of international security assistance has further destabilized Niger, making it difficult to manage the influx of climate refugees seeking shelter in neighboring countries. As political tensions rise, the situation underscores the interconnectedness of political instability and environmental crises in exacerbating the plight of displaced populations. Gaza has become the deadliest place for civilians, with ongoing violence and humanitarian crises compounding its suffering. The international community's failure to intervene or provide meaningful support has allowed the genocide to continue unabated. Despite reports and evidence of atrocities, global entities remain largely passive, with their inaction contributing to the ongoing suffering. This lack of decisive action highlights a troubling apathy toward human rights violations and underscores the urgent need for collective responsibility. The relentless conflict has resulted in widespread destruction, leaving many without homes, access to essential services, or hope for a peaceful future. International intervention and support are urgently needed to address the humanitarian needs and work toward a lasting resolution. Yemen's conflict has continued into 2024, resulting in devastating effects on its civilian population and infrastructure. The Syrian civil war has continued for years, causing immense suffering and displacement.[14] Myanmar's ongoing violence has also escalated, leading to more human rights violations and further insta-bility.[15] Now we can only wonder how these wars directly impact climate change in these regions and how it is drastically increasing the numbers of climate refugees coming from such Lands. The devastation caused by these conflicts not only exacerbates humanitarian crises but disrupts eco-systems, leading to environmental degradation. War-torn regions often see a decline in sustainable practices, further contributing to climate change and making these areas less habitable. As a result, the number of climate refugees is rising, with populations fleeing both violence and environ-mental collapse, seeking refuge in more stable regions.

It is difficult emotionally and spiritually to not let these conflicts impact us as Indigenous Peoples. This is because we know the trauma wars leave behind. These impacts are also not temporary as they continue for gener-ations until healing is achieved. For example, we commonly say that our mental health is affected by our environments and historical trauma. Our ancestors' violence and displacement still echo in our communities today, affecting our well-being. This intergenerational trauma manifests in vari-ous ways, reminding us of healing and resilience. Personally, as I got older, my life in Los Angeles exacerbated my anxiety, leading to panic disor-der. The fast-paced environment and constant noise made it difficult for me to find peace and ground myself. It wasn't until I met an Indigenous therapist that I started reflecting on what the sensations were that would come with these random panic attacks. With her guidance, I recognized the deep-seated connections between my anxiety and my father's histor-ical trauma. When I get a panic attack, I can hear loud noises like bombs exploding in the distance. This is a haunting echo of battles my father and relatives faced as children. I am transported to a space where the past and present collide, reminding me of historical violence's enduring impact. Red also overtakes my vision and hinders my ability to see. This visual element reflects the bloodshed and suffering that my father and relatives endured. It's as if the memories of their struggles are etched into my very being, forcing me to confront the lingering pain. Clinical psychologists diagnose this as post-traumatic stress disorder (PTSD). However, this is not PTSD based on my own lived experiences but rather my father's. This inherited trauma is known as intergenerational trauma, where the emotional wounds of one generation profoundly affect subsequent ones. Understanding this distinction has been crucial in my healing journey. It allows me to separate my own experiences from those of my ancestors while honoring their impact on me. I now understand that my symptoms are not just random occurrences but are deeply tied to unresolved traumas passed down through generations.

I carry a heavy heart while writing this book, especially on October 7, 2024.[16] Today marks a year since the genocide in Palestine started, a tragedy that has left deep scars on countless lives. I can now see clearly what my father and my people experienced during the genocide that also took place in our Lands. While the seas separate us, what people are experiencing in Palestine is the same thing we did a couple of years ago. This is a genocide against the Indigenous Peoples of Palestine, and we cannot ignore it.[17] It is a stark reminder that Indigenous Peoples' struggle for survival and dignity transcends borders and time. The targeted violence and systemic oppression faced by Indigenous communities is a shared history that continues to unfold in various parts of the world. Our commitment to justice and recognizing Indigenous rights must be unwavering. Indigenous Peoples are often forgotten and placed on the margins. We must actively work to dismantle the systems that perpetuate these injustices and strive to create a future where their rights are respected and upheld. This includes amplifying their voices, supporting their movements, and holding those in power accountable for their actions. Our solidarity must be more than words. It must translate into tangible efforts to ensure their survival, dignity, and self-determination. We need to go beyond acknowledgment and move collectively toward action. Many use decolonization within their rhetoric but forget decolonization is a verb, an action, and a life-long journey. Decolonization is not just a concept to discuss; it demands concrete actions that challenge and dismantle oppressive structures. It involves actively engaging in initiatives that return the Land and resources to Indigenous Peoples. To truly decolonize we must commit to transformative change and prioritize Indigenous sovereignty.

When considering what is happening in Palestine, I wonder how I would react if my community were to start a genocide. Would I have the courage to speak out against the atrocities, or would I be silent? Would I allow myself to be swayed by narratives that the government will use to justify violence under the guise of retribution? I ask these questions to understand why people are supporting this genocide. For those of us who deeply understand genocide, it is difficult to justify such actions. It is difficult to witness children's suffering. It is hard to see children feel unloved and unwanted as the world does nothing to stop their suffering. Being a community that has experienced genocide in the past is not a justification to enact genocide on another community. We live in a parallel universe where the oppressed are becoming the oppressors. The rest of the world is too afraid to speak up for fear of having negative labels placed on them. As someone who still endures the trauma that comes from her father having survived genocide, I can say that I support the Indigenous Peoples of Palestine. The memories of what my father endured as a child surviving a genocide still haunt me. These generational memories fuel my determination to share my sentiments in this discourse. By shedding light on the injustice of the past and the present, we can work toward a more just and peaceful future.

Climate justice and world peace are interconnected because environmental degradation often exacerbates social and economic inequalities, leading to conflicts over resources and Land. As climate change impacts vulnerable communities, competition for diminishing resources such as water and arable Land can fuel ongoing tensions and unrest. Thus, achieving world peace is crucial to fostering international collaboration and cooperation to implement effective climate solutions. This interconnection between climate justice and world peace extends to the displacement of populations due to climate-induced disasters, which can create humanitarian crises and strain international relations.

In times of war, environmental protection often takes a backseat, leading to further climate degradation. We are not truly advocating for climate justice if we turn a blind eye to genocides or wars, as these atrocities often arise from or contribute to resource scarcity exacerbated by climate change. We need a unified approach that prioritizes human rights and environmental protection to address these issues. We can achieve a sustainable and peaceful future by addressing these interconnected challenges. All those who stand for climate justice must also stand in solidarity with those affected by conflict in regions such as Palestine, Niger, Yemen, and Syria since these areas are often the most vulnerable to climate change impacts. Climate justice must also include a resolute stand against war and violence.

Even in times of war and conflict, Indigenous Peoples are tirelessly working to preserve a piece of their Lands. My father, for instance, always told me stories about how the guerrillas protected their Land. My father would describe how the community came together to protect sacred sites and maintain traditional practices. Despite the chaos, they cultivated the Land, passed down knowledge, and held secret ceremonies. They demonstrated unequivocal commitment to preserving their culture and Land through such resilience. This tends to be the common narrative for Indigenous Peoples when they are facing the worst of humanity. They still find solace in their Lands and environments. This demonstrates our strong understanding of how our Land and natural resources are more than commodities Western society teaches us to see them as. They are part of who we are; nature is us, and we are nature. Due to the demand for such natural resources, we are forcibly displaced and even experience war because other countries seek to obtain and control them.

Although I was young when the United States decided to invade Iraq in 2003, I heard from my community that the main reason was because of oil. For the US government, the control of these oil reserves was viewed as crucial to maintaining national security and world economic stability. As a result of this strategic interest, the US prioritized military and political influence in the region at the expense of local populations and sovereignty. Many saw the invasion of Iraq as a continuation of this policy, aimed at protecting and managing vital energy resources. I remember the somber tone in President George W. Bush's voice when he announced that US forces had begun an operation in Iraq.[18] Around the world, communities were stricken with fear and uncertainty after hearing the news. This had profound implications for global politics and the region. For those of us who deeply knew what US interference meant, we quickly just took a deep sigh and closed our eyes to remember our own relatives and community relatives we had lost. We could feel the weight of history repeating itself, as the cycle of conflict and displacement loomed. It was a moment filled with both sorrow and resilience, reminding us of the importance of holding on to our stories and traditions in the face of adversity. Every Indigenous person who has experienced war, armed conflict, or genocide never wants war to happen again. The memories of past traumas resurface, and the hope for peace becomes a distant dream. Each new conflict brings a renewed sense of dread and a longing for a future free from violence and exploitation for all people. Despite the challenges, Indigenous Peoples continue to hold on to their hopes and dreams, striving for a world where no other Indigenous community will ever have to lament their loved ones.

As a result, we as Indigenous Peoples know that we cannot advocate for the protection of our Lands, our cultures, or our people without also advocating for world peace. We understand that the fight for our rights is intrinsically linked to the broader struggle for global harmony. Our Lands and cultures can only truly thrive in a world where peace prevails, allowing us to nurture our heritage and live without fear. Therefore, advocating for world peace is not just a choice but a necessity for ensuring the safety and continuity of Indigenous communities everywhere. The United States has a long history of intervening in global conflicts, often aligning with oppressive regimes that threaten the sovereignty and well-being of Indigenous Peoples. It perpetuates cycles of violence and displacement among Indigenous communities because of this pattern of interference. Recognizing and challenging these actions are essential if a more peaceful and just world is to be achieved. For those of us who have experienced such violent atrocities in our families, the cries of mothers weeping over their children's bodies, the cries of elders crying over the loss of their homes, and the cries of scared children never leave our memories; it lingers. Seeking safety becomes the only option many have for survival. As an Indigenous person who was displaced because of climate change coupled with war, a genocide that took place against my people, I cannot write a book about climate justice without mentioning what it means to advocate for climate justice. We must seek peace; we must advocate for the end of US economic support to enact wars on these and other Indigenous Lands.

Being a climate justice advocate is not just advocating for more recycling or driving an electric car. If the same electric car companies fund the manufacturing of bombs, how are we holding them accountable? We must not allow our governments to make decisions to invade Lands because of their oil interests. Fossil fuels are not the solution to climate justice, but neither are some of the renewable energy practices being pushed forward, especially if they are leading to the desecration of Indigenous Lands. For instance, the United States removed its troops from Iraq in 2009, six years after the country decided to invade Iraq. Of course, there are oppressions that countries should ratify and eradicate, but oftentimes, as a nation, we are quick to judge others and point out their flaws instead of looking within our backyard. Throughout the Americas, Indigenous Peoples are subjected to oppression and marginalization and rarely is anything done to aid us or support our communities. Countless human rights are violated in the US every day, yet we try to use human rights violations to justify harmful interference such as wars, armed conflicts, and genocide in other countries.

It is not surprising that human rights violations occur because of climate change and wars. By displacing populations, disrupting essential services, and increasing the risk of violence and exploitation, war often exacerbates existing human rights violations. Civilians are frequently killed, tortured, and forced to flee conflict zones, where lawlessness makes such abuses possible. In addition, war can prevent humanitarian aid from reaching those in need, causing them further suffering. Likewise, climate change threatens human rights by intensifying natural disasters that result in death, destruction of homes, and loss of livelihoods. As a result of these impacts, vulnerable communities often face increased food insecurity, water scarcity, and health risks. Conflict over resources can result from worsening environments, further endangering human rights and well-being. Thus, in addition to ensuring that we do not forget the role of wars within the climate justice discourse, human rights must also be included in discussions and solutions. Part of climate policies should protect human rights as a central component to ensure the protection of vulnerable populations from both environmental and conflict-related threats. All communities can benefit from more equitable and sustainable climate action when human rights are integrated into it.

Amid all the wars, conflicts, and human rights violations in 2024, we must not forget the Indigenous children who are being displaced. Our priority should be ensuring they can access cultural teachers along their identity journeys, despite being far from their ancestral Lands. Cultural teachers can assist displaced Indigenous children with maintaining a sense of belonging and understanding their situation. Those of us who grew up in the diaspora know this is needed. This support would have provided the knowledge and connection we so desperately needed. These resources would have helped us cope with feeling lost in an unfamiliar setting. Displacement may lead to the loss of traditions, languages, and customs. Cultural teachers preserve them. They help children develop a sense of identity and self-worth by guiding and educating them on cultural practices. Cultural teachers help children transition into new environments and embrace their cultures by instilling a sense of belonging while bridging the past and present.

As I grow older, I am more aware of my role as a cultural teacher. As someone who has managed to stay connected to her roots despite being displaced, I acknowledge my privileges. I have been blessed to grow up in my maternal Indigenous community while also being able to pursue higher education. As a papaya tree that has been displaced, I have been able to adapt and thrive in new environments. In the same way that the papaya tree grows despite being uprooted, I strive to share the fruits of my experiences and knowledge with others. In many ways, my journey is a testament to resilience and the importance of preserving culture. It is important to protect our Indigenous cultures from misinformation and false claims that spread in the diaspora. Not everyone from Latin America is Indigenous, a fact underscored by the ongoing marginalization and challenges faced by Indigenous communities throughout the region. Recognizing these distinctions is vital to ensuring that we honor and accurately represent the diverse cultural narratives that exist within Latin America. Everyone has a different journey, and displacement may have led to some cultural loss. However, false claims are harmful to our diasporic communities and back in our Lands.

It is paramount that we advocate for climate justice to protect our roots both in the diaspora and back in our homelands. Like papaya trees, which rely on a delicate balance of sunlight, water, and nutrients, our communities depend on a stable climate to thrive. Climate change threatens to disrupt this equilibrium, affecting agriculture, water resources, and ultimately our cultural heritage. As we advocate for climate justice, we can ensure that these vital connections are preserved for future generations. Growing papaya trees requires careful attention to their environmental needs. To produce healthy fruit, they need well-drained soil, plenty of sunlight, and regular watering. As we nurture these trees, we can draw parallels with how we care for our global environment to support our community's growth and well-being. Like papaya trees we demonstrate the power of collective action when we are in community. Papaya trees can benefit from improved pollination and fruit production when planted in groups. This mirrors the strength of communities working together toward a common goal. We can achieve greater success in addressing climate challenges and securing a sustainable future by fostering collaboration and unity. Papaya trees grow only with the help of pollinators, which facilitate the transfer of pollen from male to female flowers. Papayas grow well when this process is carried out; it ensures the fertilization necessary for healthy development. Our efforts to attract pollinators such as bees and butterflies can boost the yield and quality of fruit, demonstrating the importance of biodiversity in agriculture. For climate justice to be achieved, Indigenous Peoples must also attract other communities willing to support us. They become our pollinators that aid our survival in the diaspora and back in our Lands.

Consequently, we must work toward Indigenous solidarity that transcends this continent, the Americas. As we build alliances across borders, we can amplify our voices and advocate more effectively for climate justice. Through these solidarities, we can share knowledge, resources, and strategies that help Indigenous communities overcome the unique challenges they face. By working together, we can build a resilient network that promotes environmental stewardship and cultural preservation. There is a need for Indigenous spaces to embrace other Indigenous communities beyond the boundaries created by settler-made borders. Creating open and welcoming Indigenous spaces fosters collaboration and mutual support among diverse communities. Through these connections, we can transcend settler-imposed borders and reinforce our commitment to cultural heritage and environmental sustainability. With Indigenous voices leading the way we can create a just and equitable world. Indigenous communities, like papaya trees, thrive when interconnected and supported by a network of allies. Like papaya trees rely on pollinators for growth, we rely on each other to nurture and sustain our communities. By embracing diverse Indigenous communities, we can enrich our cultural tapestry by learning from their experiences and practices. The diversity of perspectives fosters innovation by bringing diverse perspectives together to address shared challenges, creating more comprehensive solutions. This kind of inclusivity strengthens our collective political influence. This enables us to defend Indigenous rights and protect the environment.

Indigenous science considers the interconnectedness of all elements in the natural world, offering a holistic perspective. It emphasizes relationships and the cumulative knowledge of ecosystems, unlike Western science, which often isolates variables for study. My metaphor is that Indigenous science looks at the entire puzzle, whereas Western science only examines one or two pieces at a time. This limits the ability of Western science to grasp the intricate dynamics of ecosystems and cultural practices. When isolated variables are the focus, crucial interactions may be overlooked, leading to incomplete or skewed understandings. A holistic perspective allows us to see the bigger picture and understand the interactions among different factors, which leads to more effective, sustainable solutions. It is easier to address root causes rather than just symptoms if we consider the relationships among various elements. Climate change, biodiversity loss, and social inequality are all multifaceted issues that require this approach. As an example, the traditional fire management practices of Indigenous Australians have been recognized for reducing wildfire risks and enhancing biodiversity. These practices maintain a balance and environmental resilience in the environment by understanding natural cycles and ecological needs. Alternatively, when we only focus on isolated variables, we risk fragmenting our understanding of complex systems. Solutions may address only one aspect of a problem while neglecting how it interconnects with others. Due to the lack of consideration of broader implications and interactions, this approach may inadvertently exacerbate issues or cause new ones.

In the fight against climate change, Indigenous science contributes significantly by understanding ecosystems and sustainable practices. We can preserve our planet more effectively and culturally by incorporating their insights and leadership into environmental strategies. Unfortunately, due to their close relationship with the environment, Indigenous women are extremely vulnerable to the impacts of climate change. This vulnerability often forces them to migrate in search of more reliable live-lihood options, disrupting their communities and traditional knowledge systems. By addressing these challenges, we can better support Indige-nous women and ensure their vital contributions to climate resilience are recognized and preserved. As a result, climate justice cannot be achieved without placing Indigenous women at the forefront of environmental decision-making and policy development. Their unique perspectives and traditional knowledge are crucial for creating effective and inclusive strat-egies that address environmental and social issues. Empowering Indige-nous women ensures a holistic approach to climate resilience that benefits entire communities and ecosystems.

When we discuss climate displacement, we cannot forget how this impacts Indigenous women. Like my mother, who was uprooted from her Land due to climate change, and many Indigenous women who face sim-ilar challenges. They often bear the brunt of environmental disruptions that threaten their homes, cultures, and traditional ways of life. This dis-placement affects their physical well-being and has profound social and cultural repercussions.

Often, in Laos, Indigenous women are forced to migrate throughout the year or permanently away from their ancestral Lands to find employ-ment. As a result of this migration, they lose the connection to their cul-tural heritage as well as to their community support systems, making them vulnerable to exploitation and discrimination. Furthermore, these women may have difficulty gaining access to resources and opportunities in unfa-miliar environments, which exacerbates the impacts of climate change. Leaving their homes also increases their vulnerability to gender-based violence, particularly human trafficking, as they find themselves in pre-carious situations without the protection of their community. Due to their unfamiliarity with new environments and limited access to support networks, they are easy targets for traffickers. This means that climate-induced displacement and gender-based violence are critical challenges that demand immediate attention and action.

Our climate justice efforts must include gender-responsive solutions that prioritize the needs and voices of Indigenous women to address the social and environmental issues we face today. Our policies and programs can help protect these women from climate change and gender-based violence by being inclusive and equitable. The United Nations has highlighted that the climate crisis disproportionately impacts marginalized groups, particularly Indigenous women, due to existing social inequalities. Understanding these disparities is crucial to developing effective climate policies that protect the most vulnerable. Since Indigenous women and girls often rely on these resources for water, food, and fuel, they are vulnerable when environmental changes disrupt their availability. As the primary caregivers in many communities, they face increased burdens when resources become scarce, limiting their ability to adapt and recover. Addressing these disproportionate impacts requires empowering women and girls through education, resource access, and leadership opportunities, especially in climate decision-making. As a child, my mother was responsible for securing water and firewood for her family. These were essential parts of her daily routine, which required her to walk long distances. As a result of climate change impacts on her water and firewood reservoirs, each month required a longer walk to find these essential resources. Once manageable distances were becoming increasingly challenging, draining her energy and time. Observations like these serve as a reminder of the urgent need for community-based solutions that mitigate the effects of climate change. She managed these responsibilities with resilience and resourcefulness, despite the physical demands and time constraints.

Considering my mother's experience, we are reminded of many Indigenous women's challenges, emphasizing the need to support their vital roles in maintaining their communities. We owe our communities' resilience to Indigenous women, who form the backbone and heart of our communities. Their perseverance and strength ensure their families' and communities' survival and well-being. When we recognize and support their contributions, we can build more resilient societies better equipped to deal with climate change's challenges. Due to the contributions of Indigenous women, our communities can maintain a wealth of traditional practices, adopt adaptive strategies, and understand our ecosystems. The knowledge and ability of our community members to adapt to changing environmental conditions are of the utmost importance for our communities' survival. Indigenous women are intrinsically connected to the Land; they can manage and conserve resources effectively. Integrating their wisdom will allow us to find holistic solutions that benefit the environment and communities. Today Indigenous women continue to face significant barriers to participation in decision-making processes. Their low representation in leadership roles limits their ability to contribute meaningfully to combating the global climate crisis and addressing gender inequality. This hinders their potential but also deprives societies of valuable insights and solutions. The intersection of gender, ethnicity, and socioeconomic factors exacerbates the vulnerability of Indigenous women in the face of climate change. These overlapping challenges compound their risk exposure, so it is crucial to address systemic inequalities that perpetuate these issues. By acknowledging and addressing intersectionality, we can create more inclusive strategies that empower Indigenous women and enhance their resilience against environmental and social adversities.

For me, the teachings of my grandmother and mother played a huge role in shaping me into the Indigenous woman I am today. Their stories and experiences instilled a deep respect for our Land and traditions in me, guiding my actions and decisions. They taught me the importance of balance and harmony with nature. These lessons are vital in addressing today's environmental challenges. To honor their legacy, I strive to ensure Indigenous women's voices are heard and valued at all levels of decision-making. As Indigenous women engage, climate policies and actions benefit from their holistic, nature-focused knowledge and leadership. Their unique perspectives allow them to develop sustainable solutions that prioritize the well-being of the community and ecological balance. This inclusive approach not only enriches policy frameworks but ensures that diverse voices contribute to global climate resilience.

It is difficult to call out systems of oppression that negatively affect Indigenous Peoples, especially Indigenous women, due to systematic power imbalances, but it is essential to confront these systems to create meaningful change. By challenging and dismantling the entrenched structures that perpetuate discrimination, we can pave the way for more equitable societies. As part of this process, we need to demonstrate courage, solidarity, and a commitment to amplifying the voices of those who have been historically marginalized. Often, Indigenous women step outside their comfort zone, only to receive negative feedback.

Sacheen Littlefeather's bravery at the 1973 Academy Awards is a poignant example of the backlash Indigenous women endure when speaking out. Despite widespread attention, she was criticized and personally attacked for her brave stand against the portrayal and treatment of Native Americans in film. The incident underscores the importance of supporting Indigenous voices in these critical conversations and the resilience required to confront systemic injustices. Unfortunately, this continues to be my and other Indigenous women's everyday experience. We face barriers and resistance when advocating for our rights and perspectives, often encountering dismissive attitudes or outright hostility. People quickly jump into being defensive and focusing on their intentions while ignoring their impacts on us and our communities. They prioritize their own intentions over the actual harm caused to Indigenous communities instead of listening and reflecting. This defensiveness hinders progress and stifles meaningful dialogue, making it challenging to address the real issues. We must shift the focus from intent to impact, fostering an environment where genuine understanding and empathy can lead to transformative change. This shift often triggers reactions of denial and resistance, similar to the backlash Sacheen Littlefeather faced in 1973. People may feel threatened or uncomfortable when confronted with the realities of systemic oppression, resulting in defensiveness or attempts to discredit those who speak out. By silencing voices that are essential to driving progress and achieving justice, the status quo is perpetuated. To make true progress, climate justice must address the systemic oppression Indigenous communities face. Many Indigenous groups are on the frontlines of climate change, experiencing its impacts firsthand and possessing valuable knowledge about sustainable development. Ignoring their voices undermines our collective efforts to combat climate change and perpetuates patterns of exclusion. Despite these challenges, we remain committed to advocating for change and uplifting our voices.

It was never my mother's wish to leave her Lands. It is a sentiment shared by many Indigenous Peoples who see their ancestral territories as integral to their identity and cultural traditions. When forced to leave their Lands, people often feel profound loss and disconnection. The displacement disrupts our way of life and leaves us with a void in our hearts. Despite the challenges, Indigenous communities continue to demonstrate resilience and strength. We actively work to preserve our languages, traditions, and knowledge, ensuring that our cultures endure and thrive. Our connection to the Land remains alive in our hearts and through efforts to reclaim and revitalize our territories, we strengthen our identity for future generations. Indigenous Peoples are not going to die. We will not let our Land die. Our hopes will never die. The spirit of those who have survived poverty and hunger is unbroken. We are survivors of colonialism and genocide. Despite all these obstacles, we are still here. We must inspire and empower Indigenous youth to lead our societies in the right direction. We have seen the power of our Indigenous youth in their collective action, and they are the future. They are the light at the end of the tunnel. This tunnel is our life's journey, filled with darkness and hope. We draw strength from the wisdom and resilience of our ancestors as we move forward. Each step leads to healing and empowerment, ensuring that our cultures and communities continue to flourish. We may not live long enough to see decolonization, but we will live long enough to find our purpose. We will live long enough to see Indigenous youth thrive. Their success represents the fulfillment of our dreams and the continuation of our legacy. We aim to lay the foundation for a brighter future by nurturing their potential and fostering leadership. In their hands, our traditions will be honored, and our voices will be heard. They pave the way for future generations. We must protect our roots by empowering Indigenous youth to lead the charge for climate justice. They hold the key to safeguarding our cultural heritage and ensuring that our voices are heard in the global conversation on environmental sustainability. By supporting their efforts, we help to secure a future where our communities can thrive in harmony with the earth. In the last several years, Indigenous youth have mobilized like never before on the issue of climate justice. They have organized rallies, led campaigns, and brought attention to the importance of traditional ecological knowledge. At both the local and international levels, their activism has been instrumental in raising awareness and bringing about meaningful change. They will carry our roots forward, ensuring that the wisdom and traditions passed down through generations are not lost. As we ripen like papayas and eventually leave this world, their leadership will continue our legacies, guiding future generations to uphold the values of environmental stewardship and cultural preservation. This intergenerational commitment is vital for a just and sustainable future.

Growing papaya trees symbolizes our commitment to nurturing life and sustaining Indigenous communities. In addition to providing nourishment, these trees are also symbols of our commitment to cultivating a sustainable future. By tending to the Land and teaching our youth the importance of agriculture, we ensure that our traditions and connection to the earth remain vibrant and strong. Growing papaya trees represents nurturing and cultivating our cultural heritage and future generations. In the same way that these trees require care, attention, and patience to bear fruit, so do our efforts to preserve and pass down our cultural traditions. We are committed to long-term growth, resilience, and the sustainability of our communities.

Papaya trees serve as a powerful symbol of climate justice. They remind us of the importance of addressing environmental issues that disproportionately affect Indigenous communities. By cultivating these trees, we advocate for sustainable practices and highlight the urgent need to protect our planet for future generations. Papaya trees, with their ability to thrive in diverse environments, embody the strength and adaptability required by displaced communities. They symbolize resilience by demonstrating how life can flourish even in unfamiliar or challenging conditions. Nurturing these trees is a reminder of their enduring spirit and capacity to rebuild and sustain their cultural roots amid adversity. Despite being away from our Lands, our Land is deeply embedded in the blood that flows through our veins. The pain of our Land flows through the tears we shed in the diaspora. No matter how far we are from our Land, we are our Land.

As long as we nurture our roots, we will continue to thrive and resist, undeterred by the storms that attempt to sway us.

My mother's yearning to return to her Land is a poignant reminder of the deep bond we share with it. Her stories of childhood, filled with the scents and sounds of her homeland, echo the rhythms of our ancestors. As she dreams of the day she will walk upon the soil of her birth, her spirit remains tethered to the Land, teaching us the importance of preserving our heritage and honoring our roots.

My favorite memory is playing with caballitos
del diablo (dragonflies). They were nature's toys,
and I often share this longing memory with my family.
I used to tie a string around them and walk around my Lands,
feeling a sense of connection and joy. Those moments remind me
of the simple pleasures and the profound bond we have with our
Land, one that transcends distance and time.

JUANA BETANZOS SANTOS

In today's political climate, what we need, beyond learning, is more compassion and love for one another, our animal and plant relatives, and our Lands. Reverend Mariann Edgar Budde emphasized this message during her sermon at the national prayer service at Washington National Cathedral on January 21, 2025, "Look with compassion on the whole human family; take away the arrogance and hatred which infect our hearts; break down the walls that separate us; unite us in bonds of love; and work through our struggle and confusion to accomplish your purposes on Earth."[19]

Compassion can drive efforts to protect and preserve the environment by fostering a deeper connection between humans and the natural world. When we empathize with the plight of endangered species and fragile ecosystems, we become more motivated to take action that mitigates harm and promotes sustainability. This empathetic approach encourages policies and practices that prioritize ecological balance and long-term well-being for all living beings, including humans. As climate change increasingly displaces communities, in particular Indigenous Peoples, our compassion must extend to those affected by rising sea levels, extreme weather events, and other environmental disruptions. We need to advocate for policies that support climate refugees, ensuring they have access to safe havens and the necessary resources to rebuild their lives. By embracing empathy and solidarity, we can create a more just and equitable world for all, not just a selective few.

Just as nurturing a papaya tree requires care, attention, and patience, we also must cultivate compassion and understanding in our communities. By investing in these values, we can bear the fruits of a more harmonious and sustainable world.

A Love Letter to Displaced Indigenous Peoples

Dear Displaced Indigenous Relative,

Your longing to return to your Lands is not ignored or forgotten. We aspire to live in a world where climate change will no longer impact our communities or force them to relocate. Unfortunately, many global leaders continue to fail our communities, but our resistance and advocacy will continue to plant seedlings that will one day grow into strong trees that will not let any strong winds or obstacles destroy them.

For those displaced to the United States, January 20, 2025, marked the beginning of a nightmare as our days have since been filled with fear. Policies have been enacted that threaten our safety and undermine our rights, making it increasingly difficult for us to feel secure in our new environment. Despite these challenges, we remain steadfast in our determination to fight for justice and reclaim the dignity that has been stripped away. We are unjustly labeled as criminals on our relatives' ancestral Lands, which have been taken from them without consent. Criminalization compounds our communities' struggles, leaving us vulnerable and unsafe. Yet, we draw strength from our heritage and continue to resist these injustices with unwavering resolve.

The rise in xenophobia has led to increased discrimination and hostility toward our Indigenous Peoples, further isolating us from the broader society. Many allies who once promised support have failed to stand with us when it matters most, revealing the fragility of their false commitments. Despite these setbacks, we continue to forge ahead, building alliances with those who truly understand our struggles and working tirelessly to dismantle the systems of oppression that seek to silence us.

Our connection to the Lands is sacred and enduring, transcending any barriers imposed by external forces. This bond is woven into the very fabric of our identity, reminding us of our resilience and the strength we derive from our ancestors. No legislation or policy can sever this intrinsic relationship, as it is an integral part of who we are and who we will continue to be as Indigenous Peoples. The wisdom passed down from our ancestors provides us with guidance and fortitude in the face of adversity. Their teachings offer us a roadmap for navigating the challenges we encounter, grounding us in values of resilience, community, and perseverance. This ancestral knowledge empowers us to confront our current struggles with courage and vision, ensuring that our fight for justice is both informed and enduring for the seven generations.

Our spirit is intertwined with the very soil of our ancestral Lands, and despite the attempts to separate us, we will always find our way back. The deep connection we hold cannot be broken by borders or policies, for it is an unyielding part of our essence. We are the Land and the Land is us, and we remain committed to reclaiming our rightful place. Through our collective work, we are growing papaya trees, a symbol of our resilience and renewal. These trees stand as a testament to our enduring connection to the Land and our ability to nurture life amid adversity. By cultivating them, we are fostering not only sustenance for our community but also hope for a future where our heritage thrives.

Sincerely,

A relative who understands the pain of displacement and the relentless pursuit of belonging.

NOTES

Prologue

1   Chiara Galli, *Precarious Protections: Unaccompanied Minors Seeking Asylum in the United States* (University of California Press, 2023).
2   Amelia Frank-Vitale, "Coyotes, Caravans, and the Central American Migrant Smuggling Continuum," *Trends in Organized Crime* 26, no. 1 (2023): 64–79.
3   Jona Huber, Ignacio Madurga-Lopez, Una Murray, Peter C. McKeown, Grazia Pacillo, Peter Laderach, and Charles Spillane, "Climate-Related Migration and the Climate-Security-Migration Nexus in the Central American Dry Corridor," *Climatic Change* 176, no. 6 (2023): 79.
4   David Hernández, "Pursuant to Deportation: Latinos and Immigrant Detention," in *Latino Studies: A 20th Anniversary Reader*, ed. Lourdes Torres and Marisa Alicea (Springer Nature Switzerland, 2024), 523–58.

Introduction

1   Xóchitl C. Chávez, "Migrating Performative Traditions: The Guelaguetza Festival in Oaxacalifornia" (PhD diss., University of California, Santa Cruz, 2013).
2   Patrick Wolfe, "Settler Colonialism and the Elimination of the Native," *Journal of Genocide Research* 8, no. 4 (2006).
3   John P. Bowes, "American Indian Removal Beyond the Removal Act," *Native American and Indigenous Studies* 1, no. 1 (2014): 66–87.
4   M. E. Jones, "The Intergenerational Legacy of Indian Residential Schools," *Demography* 61, no. 6 (2024): 1871–95.
5   Randall Abate and Elizabeth Ann Kronk, eds., *Climate Change and Indigenous Peoples: The Search for Legal Remedies* (Edward Elgar Publishing, 2013).
6   Lekha Kalra, T. N. Srinatha, G. J. Abhishek, Popavath Bhargav Naik, G. S. Sujatha, Shreya S. Hanji, M. Shankar, and Pavan Kumar Kumawat, "A Comprehensive

Review of Indigenous Knowledge Systems in India and Its Importance and Role in Biodiversity Conservation," *International Journal of Environment and Climate Change* 14, no. 9 (2024): 250–65.

Chapter 1: Preparing the Soil: Our Displacement

1 William M. LeoGrande, *Our Own Backyard: The United States in Central America, 1977–1992* (University of North Carolina Press, 1998).

2 Gabriela Kovats Sánchez, "Reaffirming Indigenous Identity: Understanding Experiences of Stigmatization and Marginalization Among Mexican Indigenous College Students," *Journal of Latinos and Education* 19, no. 1 (2020): 31–44.

3 Claudio Poblete Ritschel, "Oaxaca Melting Pot of Food Cultures," *Voices of Mexico*, Autumn–Winter 2019, 80–85.

4 Kenny Cupers, "The Urbanism of Los Angeles Street Vending," in *Street Vending in the Neoliberal City*, ed. Kristina Graaff and Noa Ha (Berghahn Books, 2015), 139–63.

5 Karen Alpuche Caceres, "The Legalization of Street Vending in Los Angeles: Exploring the Impact on Vendors and Their Livelihoods" (senior thesis, Pomona College, 2019).

6 Jérémie Gilbert, *Land Grabbing, Investments & Indigenous Peoples' Rights to Land and Natural Resources* (IWGIA Report 26, 2017).

7 Silvia Bautista-Baños, Dharini Sivakumar, Arturo Bello-Pérez, Ramón Villanueva-Arce, and Mónica Hernández-López, "A Review of the Management Alternatives for Controlling Fungi on Papaya Fruit During the Postharvest Supply Chain," *Crop Protection* 49 (2013): 8–20.

Chapter 2: Uprooting Our Roots: Climate Change

1 Allan Greer, "Commons and Enclosure in the Colonization of North America," *The American Historical Review* 117, no. 2 (2012): 365–86.

2 Paulinus C. Aju, John J. I. Iwuchukwu, and Colman C. Ibe, "Our Forests, Our Environment, Our Sustainable Livelihoods," *European Journal of Academic Essays* 2, no. 4 (2015): 6–19.

3 Maarten K. Van Aalst, "The Impacts of Climate Change on the Risk of Natural Disasters," *Disasters* 30, no. 1 (2006): 5–18.

4 Jonathan Patz, Carlos Corvalan, Pierre Horwitz, Diarmid Campbell-Lendrum, Nick Watts, M. Maiero, S. Olson, et al., *Our Planet, Our Health, Our Future: Human Health and the Rio Conventions: Biological Diversity, Climate Change and Desertification* (discussion paper, 2012).

5   Juliane Zeidler and Kalemani Jo Mulongoy, "The Dry and Sub-Humid Lands Pro-
     gramme of Work of the Convention on Biological Diversity: Connecting the CBD
     and the UN Convention to Combat Desertification," *Review of European, Compara-
     tive & International Environmental Law* 12 (2003): 164.

6   William Donlan and Junghee Lee, "Indigenous and Mestizo Mexican Migrant
     Farmworkers: A Comparative Mental Health Analysis," Portland State University
     School of Social Work Faculty Publications and Presentations, 83 (2010).

7   Jorge Coronel-Bautista, "Exploring Labor Dynamics: Indigenous Identity, Wage
     Differentials, and Poverty Risk Among Mexican-Born Farmworkers in the US
     Agricultural Sector" (master's thesis, Colorado State University, 2024).

8   *Indigenous Agricultural Workers* (National Center for Farmworker Health [NCFH],
     2021), https://www.ncfh.org/uploads/3/8/6/8/38685499/indigenous_fact_sheet
     _final_12_20_2021.pdf.

9   Arcadio Viveros Guzman, "Latino Migrant Farmworkers in Saskatchewan: Occu-
     pational Health and Safety Education and the Sustainability of Agriculture" (PhD
     diss., University of Saskatchewan, 2016).

10   Heather E. Riden, Emily Felt, and Kent E. Pinkerton, "The Impact of Climate
     Change and Extreme Weather Conditions on Agricultural Health and Safety in
     California," in *Climate Change and Global Public Health*, 2nd ed., ed. Kent E. Pinker-
     ton and William N. Rom (Humana Press, 2021), 353–368.

11   Daniel E. Harmon, *Washington: Past and Present* (The Rosen Publishing Group,
     Inc, 2009).

12   Orly Stampfer, Elena Austin, Terry Ganuelas, Tremain Fiander, Edmund Seto, and
     Catherine J. Karr, "Use of Low-Cost PM Monitors and a Multi-Wavelength Aetha-
     lometer to Characterize PM2.5 in the Yakama Nation Reservation," *Atmospheric
     Environment* 224 (2020): 117292.

13   Moussa El Khayat, Dana A. Halwani, Layal Hneiny, Ibrahim Alameddine, Musta-
     pha A. Haidar, and Rima R. Habib, "Impacts of Climate Change and Heat Stress on
     Farmworkers' Health: A Scoping Review," *Frontiers in Public Health* 10 (2022): 782811.

14   Julie Sabatier, "Oregon Has New Rules to Protect Workers from Extreme Heat, but
     Advocates Worry About Enforcement," *OPB*, July 21, 2021, https://www.opb.org
     /article/2021/07/09/oregon-new-rules-workers-protections-extreme-heat-wave
     -advocates-enforcement/.

15   Leila Duntley, "WASHINGTON: Updated Heat Illness Rules Effective July 17,
     2023," *Vigilant*, July 6, 2023, https://www.vigilant.org/employment-law-blog
     /washington-updated-heat-illness-rules-effective-july-17-2023/.

16   Jennifer O'Rourke, "The Overlooked Communities of Forced Displacement in the
     United States: Humanizing the Relocation of Indigenous Tribes in the Face of Cli-
     mate Change," *University of Cincinnati Law Review* 92 (2023): 850.

17   Bethuel Sibongiseni Ngcamu, "Climate Change Effects on Vulnerable Populations in the Global South: A Systematic Review," *Natural Hazards* 118, no. 2 (2023): 977–91.

18   Alex Latu, "Pacific Islands Review (2021–24)," July 12, 2024, https://ssrn.com/abstract=4892543.

19   Adam B. Lerner and Ben O'Loughlin, "Strategic Ontologies: Narrative and Meso-Level Theorizing in International Politics," *International Studies Quarterly* 67, no. 3 (2023).

20   Edwin A. Hernández-Delgado, "Coastal Restoration Challenges and Strategies for Small Island Developing States in the Face of Sea Level Rise and Climate Change," *Coasts* 4, no. 2 (2024): 235–86.

21   Soseala Saosaoa Tinilau, "An Examination of Formal, Non-formal, and Informal Environmental Stewardship Education in Tuvalu" (PhD diss., University of Lincoln, 2024).

22   Sarah Kuta, "Mexico's Howler Monkeys Are Dying, 'Falling Out of the Trees,' Amid Scorching Heat Wave," *Smithsonian* magazine, May 22, 2024, https://www.smithsonianmag.com/smart-news/mexicos-howler-monkeys-are-dying-falling-out-of-the-trees-amid-scorching-heat-wave-180984403/.

23   Brett Christophers, "The End of Carbon Capitalism (as We Knew It)," *Critical Historical Studies* 8, no. 2 (2021): 239–69.

24   Tim Di Muzio, *Carbon Capitalism: Energy, Social Reproduction and World Order* (Rowman & Littlefield, 2015).

25   Rabindra Garada, "Dynamics of Coal Mining Caused Environmental Crisis Versus Displaced People's Question of Survival: A Case of Talcher Coal Belt, Odisha (India)," *Global Journal of Human Social Science, Geography, Geo-Sciences, Environmental Disaster Management* 13, no. 6 (2013).

26   Peter Newell and Matthew Paterson, "Climate Capitalism," in *After Cancún: Climate Governance or Climate Conflicts*, ed. Elmar Altvater and Achim Brunnengräber (VS Verlag für Sozialwissenschaften, 2011), 23–44.

27   Naomi Klein, *This Changes Everything: Capitalism vs. the Climate* (Simon and Schuster, 2015).

28   Raphael Calel, "Carbon Markets: A Historical Overview," *Wiley Interdisciplinary Reviews: Climate Change* 4, no. 2 (2013): 107–19.

29   Teresa De La Fuente and Reem Hajjar, "Do Current Forest Carbon Standards Include Adequate Requirements to Ensure Indigenous Peoples' Rights in REDD Projects?," *International Forestry Review* 15, no. 4 (2013): 427–41.

30   Carlandio Alves da Silva, Sheila Castro dos Santos, and Onelia Carmem Rossetto, "The Paiter Suruí Indigenous Peoples in Defence of Their Territory: The Case of the Suruí Forest Carbon Project (PCFS)—RONDONIA/BRAZIL," in *Traditional Knowledge and Climate Change: An Environmental Impact on Landscape and Communities*, ed. Ana Penteado, Shambhu Prasad Chakrabarty, and Owais H. Shaikh (Springer Nature Singapore, 2024), 111–32.

31   Guineverre Alvarez, Maria Elfving, and Célio Andrade, "REDD+ Governance and
      Indigenous Peoples in Latin America: The Case of Suru Carbon Project in the Bra-
      zilian Amazon Forest," *Latin American Journal of Management for Sustainable Devel-
      opment* 3, no. 2 (2016): 133–46.

32   Chris Van Dam, "Indigenous Territories and REDD in Latin America: Opportunity
      or Threat?," *Forests* 2, no. 1 (2011): 394–414.

33   da Silva, Castro dos Santos, and Rossetto, "The Paiter Suruí Indigenous Peoples in
      Defence of Their Territory," 111–32.

34   Juliana Luna Freire, "Epistemological Spaces, Carbon Credits, and Environmental
      Modernity: The Suruí Forest Carbon Project," *TRANSMODERNITY: Journal of
      Peripheral Cultural Production of the Luso-Hispanic World* 7, no. 2 (2017).

35   Sumeet Jhamb and Ryan Navrot, "Decarbonizing the Global Economy: A Carbon
      Credits Analysis of Alaska Native Corporations," *IUP Journal of Business Strategy* 19,
      no. 4 (2022): 26–38.

36   Sarah L. Schooler, "Weather, Brown Bears and Timber: Direct and Indirect Drivers
      of a Northern Ungulate Population" (PhD diss., SUNY College of Environmental
      Science and Forestry, 2022).

37   Jhamb and Navrot, "Decarbonizing the Global Economy," 26–38.

38   Kathleen Birrell, Lee Godden, and Maureen Tehan, "Climate Change and REDD+:
      Property as a Prism for Conceiving Indigenous Peoples' Engagement," *Journal of
      Human Rights and the Environment* 3, no. 2 (2012): 196–216.

39   Shannan L. Mattiace, *To See with Two Eyes: Peasant Activism and Indian Autonomy in
      Chiapas, Mexico* (UNM Press, 2003).

40   Tatiana Pérez Ramírez, "Zapatistas, Anti-Zapatistas and Other Approaches: A His-
      toriographic Review of the Revolution in Mexico State," *Historia Mexicana* 73,
      no. 4 (2024).

41   Marlon Vladimir Escamilla Rodríguez, "Nahua-Pipil Diasporic Migration and
      Symbolic Landscape in Early Postclassic El Salvador" (PhD diss., Vanderbilt Uni-
      versity, 2022).

42   Frank Biermann and Ingrid Boas, "Protecting Climate Refugees: The Case for a
      Global Protocol," *Environment: Science and Policy for Sustainable Development* 50,
      no. 6 (2008): 8–17.

43   Anna R. Welch and Sara P. Cressey, "Dropping the Veil: How an Investigation into
      One Asylum Office Reveals Systemic Failures Within the US Affirmative Asylum
      System," *Loyola of Los Angeles Law Review* 57, no. 1 (2024).

44   Ryan Baugh, *Refugees and Asylees: 2021* (Department of Homeland Security Office of
      Immigration Statistics, Annual Flow Report, September 2022), accessed June 13, 2024.

45   Diana M. Liverman, "Vulnerability and Adaptation to Drought in Mexico," *Natural
      Resources Journal* 39 (1999): 99.

Chapter 3: Preserving Our Land: Land Rights

1   Fulvio Mazzocchi, "Analyzing Knowledge as Part of a Cultural Framework: The Case of Traditional Ecological Knowledge," *Environments: A Journal of Interdisciplinary Studies* 36, no. 2 (2008).

2   Jo-Ann Archibald, *Indigenous Storywork: Educating the Heart, Mind, Body, and Spirit* (UBC Press, 2008).

3   Hovan T. Lawton, "Central American Saints: The Formation and Preservation of Latter-Day Saint Community and Identity in El Salvador and Guatemala, 1960–1992" (master's thesis, Utah State University, 2023).

4   Chris Goertzen, *Made in Mexico: Tradition, Tourism, and Political Ferment in Oaxaca* (University Press of Mississippi, 2010).

5   Julio Gutiérrez, "Real Estate Oligarchs: Elites and the Urbanization of the Land Question in El Salvador," *The Journal of Peasant Studies* 51, no. 2 (2024): 489–511.

6   Ruchi Patel, "Securing Development: Uneven Geographies of Coastal Tourism Development in El Salvador," *World Development* 174 (2024): 106450.

7   K. Taiuru, "Treaty of Waitangi/Te Tiriti and Māori Ethics Guidelines for: AI, Algorithms, Data and IOT," *Te Kete o Karaitiana Taiuru* (blog), May 4, 2020, http://www.taiuru.Maori.nz/TiritiEthicalGuide.

8   Dominic O'Sullivan, Heather Came, Tim McCreanor, and Jacquie Kidd, "A Critical Review of the Cabinet Circular on Te Tiriti o Waitangi and the Treaty of Waitangi Advice to Ministers," *Ethnicities* 21, no. 6, (2021): 1093–112.

9   Margaret Mutu, "Māori Issues," *The Contemporary Pacific* 32, no. 1 (2020): 240–49.

10 Luis Castro Castro and Daniuska González González, "El 'diablo del amor': La demonización de los afectos, la coquetería y la carnalidad femenina en el virreinato del Perú y la capitanía general de Chile (siglos XVI–XVIII)," *Confluenze. Rivista Di Studi Iberoamericani* 12, no. 1 (July 24, 2020): 512–43.

11 Iliana Monterroso, "How Can Tenure Reform Processes Lead to Community-Based Resource Management? Experiences from Latin America," in *Routledge Handbook of Latin America and the Environment*, ed. Beatriz Bustos et al. (Routledge, 2023), 335–49.

12 Soukphavanh Sawathvong and Kimihiko Hyakumura, "A Comparison of the Free, Prior, and Informed Consent (FPIC) Guidelines and the 'Implementation of Governance, Forest Landscapes, and Livelihoods' Project in Lao PDR: The FPIC Team Composition and the Implementation Process," *Land* 13, no. 4 (2024): 408.

13 Dorothée Cambou, "The UNDRIP and the Legal Significance of the Right of Indigenous Peoples to Self-Determination: A Human Rights Approach with a Multidimensional Perspective," in *The United Nations Declaration on the Rights of Indigenous Peoples*, ed. Damien Short et al. (Routledge, 2020), 33–49.

14   Antonietta Di Blase and Valentina Vadi, "Introduction," in *The Inherent Rights of Indigenous Peoples in International Law* (RomaTrePress, 2020), 15–39.

15   Martin Papillon, Jean Leclair, and Dominique Leydet, "Free, Prior and Informed Consent: Between Legal Ambiguity and Political Agency," *International Journal on Minority and Group Rights* 27, no. 2 (2020): 223–32.

16   Alessandra Bergamin, "The Death Squads Hunting Environmental Defenders," *In These Times* 48, no. 5 (2024).

17   C. Gordon, "Criminalizing Care: Environmental Justice Under Political and Police Repression," *Environmental Communication* 18, no. 1–2 (2024): 138–45.

18   Philippe Le Billon and Päivi Lujala, "Environmental and Land Defenders: Global Patterns and Determinants of Repression," *Global Environmental Change* 65 (2020): 102163.

19   Christopher Loperena, "Frontiers of Dispossession, Territories of Freedom," *NACLA Report on the America* 53, no. 3 (2021): 211–14.

20   Global Witness, "More than 2,100 Land and Environmental Defenders Killed Globally between 2012 and 2023," press release, September 10, 2024, https://www .globalwitness.org/en/press-releases/more-2100-land-and-environmental-defenders -killed-globally-between-2012-and-2023/.

21   Peter Veit, "Land Matters: How Securing Community Land Rights Can Slow Climate Change and Accelerate the Sustainable Development Goals," *Global Witness*, January 24, 2019, accessed September 22, 2024, https://www.wri.org/insights/land -matters-how-securing-community-land-rights-can-slow-climate-change-and -accelerate#.

22   Deborah Carvalho, "Patriarchy, Culture and Land: Challenges in Securing Women's Ownership and Titling Rights in La Paz, Bolivia" (master's thesis, Simon Fraser University, 2012).

23   Hama Arba Diallo, "United Nations Convention to Combat Desertification (UNCCD)," in *The Future of Drylands: International Scientific Conference on Desertification and Drylands Research Tunis, Tunisia, 19–21 June 2006*, ed. Cathy Lee and Thomas Schaaf (Springer Netherlands, 2008), 13–16.

24   Steven A. Kennett and Arlene J. Kwasniak, "Property Rights and the Legal Framework for Carbon Sequestration on Agricultural Land," *Ottawa Law Review* 37 (2005): 171.

25   Alessandro Barghini and Marney Pascoli Cereda, "Traditional Amerindian Cassava-Based Foods and Drinks in the Amazon Region," in *Traditional Starch Food Products*, ed. Marney Pascoli Cereda and Olivier François Vilpoux (Academic Press, 2025), 129–54.

26   Mario L. Cardozo, Danilo Salas, Isabel Ferreira, Teresa Mereles, and Laura Rodríguez, "Soy Expansion and the Absent State: Indigenous and Peasant Livelihood Options in Eastern Paraguay," *Journal of Latin American Geography* 15, no. 3 (2016): 87–104.

27 Héctor A. Keller and Verónica L. Lozano, "Socioecological Impacts of Pine Mono-
    cultures on Guaraní Territories in Argentina: The Hidden Costs of Modern Devel-
    opment," *Inland Waters* (2024): 1–35.

28 Lía Rodríguez de la Vega and Héctor Rodríguez, "Quality of Life of the Guaraní
    Community," in *Quality of Life in Communities of Latin Countries*, ed. Graciela Tonon
    (Springer Cham, 2017), 137–65.

29 Scott Allen and Pamela McPherson, "We Warned DHS That a Migrant Child
    Could Die in US Custody. Now One Has," *The Washington Post*, December 19, 2018,
    https://www.washingtonpost.com/outlook/2018/12/19/we-warned-dhs-that
    -migrant-child-could-die-us-custody-now-one-has/.

30 Jeremy Slack, Daniel E. Martínez, and Josiah Heyman, *Immigration Authorities Sys-
    tematically Deny Medical Care for Migrants Who Speak Indigenous Languages* (Center
    for Migration Studies, December 2018).

31 Greg Grandin and Elizabeth Oglesby, "Washington Trained Guatemala's Killers for
    Decades," *The Nation*, January 25, 2019, https://www.thenation.com/article/archive
    /border-patrol-guatemala-dictatorship/.

32 Autumn Knowlton, "Q'eqchi' Mayas and Defense of Territory: Learning Through
    the Contentious Politics of Land in 'Post-Conflict' Guatemala" (PhD diss., Univer-
    sity of British Columbia, 2016).

33 Liza Grandia, *Enclosed: Conservation, Cattle, and Commerce Among the Q'eqchi' Maya
    Lowlanders* (University of Washington Press, 2012).

34 Sammy Westfall, Brian Murphy, Adam Taylor, and Bryan Pietsch, "The Israeli-
    Palestinian Conflict: A Chronology," *The Washington Post*, November 6, 2023,
    https://www.washingtonpost.com/world/2023/israel-palestine-conflict-timeline
    -history-explained/.

35 Andrew L. Gulley, "China, the Democratic Republic of the Congo, and Artisanal
    Cobalt Mining from 2000 Through 2020," *Proceedings of the National Academy of
    Sciences* 120, no. 26 (2023): e2212037120.

36 Mohammed Nijim, "Genocide in Palestine: Gaza as a Case Study," *The International
    Journal of Human Rights* 27, no. 1 (2023): 165–200.

37 Brendan Ciarán Browne, "Pursuing International Criminal Justice, the ICC, and
    Palestine," in *Transitional (in)Justice and Enforcing the Peace on Palestine* (Springer
    International Publishing, 2023), 61–78.

38 "Gaza: Hamas, Israel Committed War Crimes, Claims Independent Rights Probe,"
    UN News, June 12, 2024, accessed September 2, 2024, https://news.un.org/en/story
    /2024/06/1150946.

39 Suha Jarrar, "Adaptation Under Occupation: Climate Change Vulnerability," in *Pro-
    longed Occupation and International Law*, ed. Nada Kiswanson and Susan Power (Brill
    Nijhoff, 2023), 176–96.

40　Stephen Uche Onyigbuo, "Stepping into the Stairs or Staring into the Steps of the Israeli-Hamas War: 7 October 2023 and the Days After," April 23, 2024, http://dx.doi.org/10.2139/ssrn.4804264.

41　Sara Salazar Hughes, Stepha Velednitsky, and Amelia Arden Green, "Greenwashing in Palestine/Israel: Settler Colonialism and Environmental Injustice in the Age of Climate Catastrophe," *Environment and Planning E: Nature and Space* 6, no. 1 (2023): 495–513.

42　Eyad Yaqoub Yaqob, "Climate Change Implication on Palestine: A Case Study Jenin Governorate," *American Journal of Multidisciplinary Research and Innovation* 2, no. 3 (2023): 70–76.

43　Erda Çeler and Yusuf Serengil, "A Multi-Scale Climate Vulnerability and Risk Assessment (C-VRA) Methodology for Corporate Scale Investments: West Bank-Palestine Case Study," *Resilience* 7, no. 2 (2023): 269–92.

44　Ornit Avidar, "Israel: From Water Scarcity to Water Surplus," in *The Palgrave International Handbook of Israel*, ed. P. R. Kumaraswamy (Springer Nature Singapore, 2023), 1–14.

45　Alon Tal, "Unkept Promises: Israel's Implementation of Its International Climate Change Commitments," *Israel Journal of Foreign Affairs* 14, no. 1 (2020): 21–51.

46　Lotte de Jong, Sophie De Bruin, Joost Knoop, and Jasper van Vliet, "Understanding Land-Use Change Conflict: A Systematic Review of Case Studies," *Journal of Land Use Science* 16, no. 3 (2021): 223–39.

47　Saturnino M. Borras and Jennifer C. Franco, "The Challenge of Locating Land-Based Climate Change Mitigation and Adaptation Politics Within a Social Justice Perspective: Towards an Idea of Agrarian Climate Justice," in *Converging Social Justice Issues and Movements*, ed. Tsegaye Moreda et al. (Routledge, 2020), 82–99.

48　Koen Vlassenroot and Chris Huggins, "Land, Migration and Conflict in Eastern DRC," in *From the Ground Up: Land Rights, Conflict and Peace in Sub-Saharan Africa*, ed. Chris Huggins and Jenny Clover (Institute for Security Studies, 2005), 115–94.

49　Filip Johnsson, Jan Kjärstad, and Johan Rootzén, "The Threat to Climate Change Mitigation Posed by the Abundance of Fossil Fuels," *Climate Policy* 19, no. 2 (2019): 258–74.

50　Kerry Black and Edward McBean, "Increased Indigenous Participation in Environmental Decision-Making," *International Indigenous Policy Journal* 7, no. 4 (2016): 1–24.

Chapter 4: Harvesting Our Present: Renewable Energy

1　Richard Oldani, "Energy Transformation and Flow," *Journal of Earth and Environmental Science Research* 5, no. 2 (2023): https://doi.org/10.47363/JEESR/2023(5)189.

2　Helen E. Longino, *Science as Social Knowledge: Values and Objectivity in Scientific Inquiry* (Princeton University Press, 1990).

3   Glen S. Aikenhead and Masakata Ogawa, "Indigenous Knowledge and Science Revisited," *Cultural Studies of Science Education* 2 (2007): 539–620.

4   Roger Fouquet and Peter J. G. Pearson, "A Thousand Years of Energy Use in the United Kingdom," *The Energy Journal* 19, no. 4 (1998): 1–41.

5   John U. Nef, "An Early Energy Crisis and Its Consequences," *Scientific American* 237, no. 5 (1977): 140–51.

6   Onur Ulas Ince, "Primitive Accumulation, New Enclosures, and Global Land Grabs: A Theoretical Intervention," *Rural Sociology* 79, no. 1 (2014): 104–31.

7   Jack A. Goldstone, "Efflorescences and Economic Growth in World History: Rethinking the 'Rise of the West' and the Industrial Revolution," *Journal of World History* (2002): 323–89.

8   N. S. Mammadov, N. A. Ganiyeva, and G. A. Aliyeva, "Role of Renewable Energy Sources in the World," *Journal of Renewable Energy, Electrical, and Computer Engineering* 2, no. 2 (2022): 63–67.

9   Alexandra Pelas, "Finding 'New Oil' in the Silver State: A Cost Benefit Analysis of Lithium Mining in Nevada," *Governance: The Political Science Journal at UNLV* 7, no. 1 (2023): 4.

10 Kelsey R. Hill, "'Bad Medicine' at Peehee Mu'huh: An Environmental History of Thacker Pass" (PhD diss., University of Nevada, Reno, 2024).

11 Jack Healy and Mike Baker, "As Miners Chase Clean-Energy Minerals, Tribes Fear a Repeat of the Past," *The New York Times*, December 27, 2021, https://www.nytimes .com/2021/12/27/us/mining-clean-energy-antimony-tribes.html.

12 Zachary Johnigan, "Federal Government: Assessing the Impact of Lithium Mineral Resource Extraction on Indigenous Communities in The State of Nevada" (capstone project, University of Nevada, Las Vegas, 2024).

13 Hill, "'Bad Medicine' at Peehee Mu'huh."

14 Buddy Borden and Tom Harris, "Estimated Economic and Fiscal Impacts from New Lithium Mining and Processing Operations in Humboldt County, Nevada" (Extension, University of Nevada, Reno, 2023).

15 Paul Feather, "Finding Ourselves at Peehee Mu'huh: An Interview with Daranda Hinkey," *CounterPunch*, June 4, 2021.

16 Maurizio Malpede, "The Dark Side of Batteries: Child Labor and Cobalt Mining in the Democratic Republic of Congo," Working Paper Series No. 22 (Centre for Research on Geography, Resources, Environment, Energy and Networks, Bocconi University, 2022).

17 John T. Williams, Achim Mambu Vangu, Habib Balu Mabiala, Honore Bambi Mangungulu, and Elizabeth K. Tissingh, "Toxicity in the Supply Chain: Cobalt, Orthopaedics, and the Democratic Republic of the Congo," *The Lancet Planetary Health* 5, no. 6 (2021): e327–e328.

18   Célestin Banza Lubaba Nkulu, Lidia Casas, Vincent Haufroid, Thierry De Putter, Nelly D. Saenen, Tony Kayembe-Kitenge, Paul Musa Obadia, et al., "Sustainability of Artisanal Mining of Cobalt in DR Congo," *Nature Sustainability* 1, no. 9 (2018): 495–504.

19   R. Hickman, "Hybrids, EVs and Greenwashing," *Town & Country Planning* 90, no. 7&8 (2021): 221–23.

20   Roos Haer, Christopher Michael Faulkner, and Beth Elise Whitaker, "Rebel Funding and Child Soldiers: Exploring the Relationship Between Natural Resources and Forcible Recruitment," *European Journal of International Relations* 26, no. 1 (2020): 236–62.

21   James Rodríguez, "Where Impunity Reigns: Nickel Mining in El Estor, Guatemala," *Latin American Perspectives* 48, no. 1 (2021): 289–99.

22   Juan Pablo Gramajo Pineda, "Propuesta de un plan para la optimización del proceso de trituración y separación de carbón mineral en la compañía procesadora de níquel de Izabal S.A. ubicada en El Estor Izabal" (Universidad Rural de Guatemala, 2022).

23   Anna G. Sveinsdóttir, Mariel Aguilar-Støen, and Benedicte Bull, "Resistance, Repression and Elite Dynamics: Unpacking Violence in the Guatemalan Mining Sector," *Geoforum* 118 (2021): 117–29.

24   Rodríguez, "Where Impunity Reigns," 289–99.

25   David Leveille, "A Guatemalan Indigenous Land Rights Activist Wins the Goldman Environmental Prize," *The World,* April 24, 2017, accessed May 2, 2024, https://theworld.org/stories/2017/04/24/guatemalan-land-activist-wins-goldman-prize.

26   Jill E. Johnston and Andrea Hricko, "Industrial Lead Poisoning in Los Angeles: Anatomy of a Public Health Failure," *Environmental Justice* 10, no. 5 (October 2017): 162–67, https://doi.org/10.1089/env.2017.0019.

27   María Elena Huesca-Pérez, Claudia Sheinbaum-Pardo, and Johann Köppel, "Social Implications of Siting Wind Energy in a Disadvantaged Region—The Case of the Isthmus of Tehuantepec, Mexico," *Renewable and Sustainable Energy Reviews* 58 (2016): 952–65.

28   Jacobo Ramirez, "Contentious Dynamics Within the Social Turbulence of Environmental (In)justice Surrounding Wind Energy Farms in Oaxaca, Mexico," *Journal of Business Ethics* 169, no. 3 (2021): 387–404.

29   María Elena Huesca-Pérez, Claudia Sheinbaum-Pardo, and Johann Köppel, "From Global to Local: Impact Assessment and Social Implications Related to Wind Energy Projects in Oaxaca, Mexico," *Impact Assessment and Project Appraisal* 36, no. 6 (2018): 479–93.

30   Alexander Dunlap and Martín Correa Arce, "'Murderous Energy' in Oaxaca, Mexico: Wind Factories, Territorial Struggle and Social Warfare," *The Journal of Peasant Studies* 49, no. 2 (2022): 455–80.

31   Samira Lobato Manrique, "Conciertos gratuitos en el Zócalo: la estrategia de cam-
     paña de Claudia Sheinbaum" (thesis, Universidad Veracruzana, August 2024).

Chapter 5: Nurturing Seedlings: Our Youth

1   Christian Paz, "The Los Angeles City Council's Racist Recording Scandal,
     Explained," *Vox*, October 2022.
2   Darragh Roche, "What Did Nury Martinez Say?," *Newsweek*, October 2022.
3   John Antczak, "Los Angeles Council President Resigns After Racist Remarks," *AP
     News*, October 2022.
4   Laura C. Chávez-Moreno, "The Problem with Latinx as a Racial Construct vis-a-
     vis Language and Bilingualism: Toward Recognizing Multiple Colonialisms in the
     Racialization of Latinidad," in *Handbook of Latinos and Education: Theory, Research,
     and Practice*, ed. Enrique G. Murillo et al. (Routledge, 2021), 250–61.
5   Laura Meneses, "Undoing Latinidad: An Intersectional Investigation into the
     Colonial Legacies of Latinx Representations" (major research paper, University of
     Ottawa, 2021).
6   Chew Márquez and Aiken Samuel, "La radio como herramienta para el ejercicio de
     los derechos: una etnografía en Muqb'ilha'II, Alta Verapaz" (PhD diss., Universidad
     del Valle de Guatemala, 2013).
7   Eric Hoenes del Pinal, "The Promises and Perils of Radio as a Medium of Faith in a
     Q'eqchi'-Maya Catholic Community," *Journal of Global Catholicism* 3, no. 2 (2019): 4.
8   Daniel R. Wildcat, *On Indigenuity: Learning the Lessons of Mother Earth* (Fulcrum
     Publishing, 2023).
9   Cristina V. Kleinert and Christiane Stallaert, "Mexican Indigenous Languages and
     Public Service Connections. An Ethnographic Decolonial Perspective," in *Critical
     Approaches to Institutional Translation and Interpreting*, ed. Esther Monzó-Nebot and
     María Lomeña-Galiano (Routledge, 2024), 97–116.
10   Benjamin Miller, *Distant Readings of Disciplinarity: Knowing and Doing in Composi-
     tion/Rhetoric Dissertations* (Utah State University Press, 2022).

Chapter 6: Protecting Our Roots: Climate Justice

1   Safdar Sial and Tanveer Anjum, "Jihad, Extremism and Radicalization: A Public
     Perspective," *Conflict and Peace Studies* 3, no. 2 (2010): 33–62.
2   zoë (@zoenone0none), "you are significantly closer to being a climate refugee than a
     billionaire," Twitter (now X), September 15, 2020, https://x.com/zoenone0none
     /status/1305935968101253122.

3   Ashfaq Khalfan, Astrid Nilsson Lewis, Carlos Aguilar, Jacqueline Persson, Max
    Lawson, Nafkote Dabi, Safa Jayoussi, and Sunil Acharya, *Climate Equality: A Planet
    for the 99%* (Oxfam International briefing paper, 2023).
4   Clinton G. Wallace and Shelley Welton, "Taxing Luxury Emissions," *Cornell Law
    Review* 109, no. 5 (2023): 1153–232.
5   Säde Hormio, "Carbon Inequality and Direct Responsibility," in *Taking Responsibil-
    ity for Climate Change* (Springer International Publishing, 2024), 101–21.
6   Khalfan et al., "Climate Equality."
7   Dario Kenner, *Carbon Inequality: The Role of the Richest in Climate Change* (Rout-
    ledge, 2019).
8   Brian Tokar, "Principles for Just and Effective Action," in *The Climate Crisis: Sci-
    ence, Impacts, Policy, Psychology, Justice, Social Movements* (Cambridge University
    Press, 2022), 189.
9   Tsegaye Moreda, "Beyond Land Rights Registration: Understanding the Mundane
    Elements of Land Conflict in Ethiopia," *The Journal of Peasant Studies* 50, no. 5
    (2023): 1791–819.
10 Philippe Pellet, "Understanding the 2020–2021 Tigray Conflict in Ethiopia—Back-
    ground, Root Causes, and Consequences," *Kki Elemzések* 2021, no. 39 (2021): 1–20.
11 Wubante Ayalew Dessie, Simeneh Bires Belete, and Agenagn Kebede Dagnew,
    "The Conflict Between the Tigray People Liberation Front (TPLF) and the Fed-
    eral Democratic Republic of Ethiopia (FDRE): Special Emphasis on the Pretoria
    Agreement," *African Security Review* 33, no. 4 (2024): 420–36.
12 Andrey Korotayev, Leonid Issaev, Anna Ilyina, Julia Zinkina, and Elena Voronina,
    "Revolutionary History of Niger: From Independence to 2023 Coup," in *Terrorism
    and Political Contention: New Perspectives on North Africa and the Sahel Region*, ed.
    János Besenyő, Leonid Issaev, and Andrey Korotayev (Springer Nature Switzerland,
    2024), 169–94.
13 Malik Olatunde Oduoye, Samuel Chinonso Ubechu, Habiba Zafar, Zainab Noor,
    Usman Abolaji Oyeleke, and Ganiyat Temitope Agbeyewo, "Humanitarian Crisis
    amid the Military Coup in Niger Republic; What Went Wrong?," *Health Science
    Reports* 7, no. 6 (2024): e2180.
14 Aleksandar Kešeljević and Rok Spruk, "Estimating the Effects of Syrian Civil War,"
    Empirical Economics 66, no. 2 (2024): 671–703.
15 Justine Chambers and Michael R. Dunford, eds., *Myanmar in Crisis: Living with the
    Pandemic and the Coup* (ISEAS-Yusof Ishak Institute, 2023).
16 Shahridan Mohd Fathil and Ina Ismiarti Shariffuddin, "Our Choices in the Face of
    Genocide: Resistance or Collaboration," *Malaysian Journal of Anaesthesiology* 3, no. 1
    (2024): 7–10.

17 Zainab Zahid, "How CNN Portrays Israel as the Victim in the Context of the Octo-
    ber 7th, 2023 Israeli-Palestinian Hostilities" (master's thesis, University of Chicago,
    2024).

18 Joy Gordon, *Invisible War: The United States and the Iraq Sanctions* (Harvard Univer-
    sity Press, 2010).

19 Siladitya Ray, "What Did the Bishop Say to Trump During the Inaugural Prayer
    Service? Here's the Full Transcript," *Forbes*, January 22, 2025, https://www.forbes
    .com/sites/siladityaray/2025/01/22/what-did-the-bishop-say-to-trump-during
    -prayer-service-heres-the-full-transcript/.

INDEX

ACKNOWLEDGMENTS

I t is an immense privilege to write my second book, and I could not have done it without everyone's support. I want to thank my very loving and dedicated parents, Juana and Victor, who have played the biggest role in shaping who I am today. Their stories are part of my being and make me the dedicated Indigenous advocate I am today. I also want to thank my brother, Victor Jr. He has taught me the role of being a sister not just to them but also to others.

Keon, my beloved partner, has been an unwavering source of inspiration and support throughout this journey. His encouragement and wisdom have empowered me to embrace my voice and share my experiences with confidence. I am deeply grateful for his presence in my life and the profound impact he has had on my personal and professional growth. I also want to express my heartfelt gratitude to my father-in-law, Ed, and my sister-in-law, Kiki. Their resilience and passion for life have been a constant inspiration to me. Their support and encouragement have been instrumental in shaping my perspective and guiding me through this journey. Also, a beautiful spiritual embrace to my mother-in-law, Joanne, who watches over me from above. Her gentle spirit and love continue to guide me on my path, providing comfort and strength in times of need.

I extend a warm embrace to my North Atlantic Books family, who have played a tremendous role in this book. Their commitment, expertise, and passion have been invaluable in bringing my vision to life. I am incredibly thankful for their belief in my work and tireless efforts to ensure this book reaches readers worldwide.

Lastly, I want to express my heartfelt gratitude to my readers. Your continuous support and engagement have been a driving force behind my writing journey. Knowing that my words resonate with you has given me the courage to keep writing and sharing my stories.

ABOUT THE AUTHOR

PHOTO BY VICTOR MANUEL HERNANDEZ JR.

JESSICA HERNANDEZ, PhD, is an Indigenous scholar, scientist, and community advocate based in the Pacific Northwest. She has an interdisciplinary academic background ranging from marine sciences to forestry. Her work is grounded in her Indigenous cultures and ways of knowing that are rooted from El Salvador (Maya Ch'orti') and Oaxaca, Mexico (Zapotec). She advocates for food, climate, and environmental justice through her scientific and community work and strongly believes that Indigenous sciences can heal our Indigenous Lands. She was raised in South Central Los Angeles, and in 2020 she became the first alum from her high school to receive and complete a doctoral degree. She is the founder of Pina Soul, SPC, an environmental consulting and artesanías hybrid business that promotes and supports environmental sustainability and conservation among Black and Indigenous communities.

About
North Atlantic Books

North Atlantic Books (NAB) is an independent nonprofit publisher committed to a bold exploration of the relationships between mind, body, spirit, and nature. Founded in 1974, NAB aims to nurture a holistic view of the arts, sciences, humanities, and healing. To make a donation or to learn more about our books, authors, events, and newsletter, please visit www.northatlanticbooks.com.